A SYSTEM OF STAGES FOR
CORRELATION OF
MAGALLANES BASIN SEDIMENTS

The Geological Society of America, Inc.
Memoir 139

A System of Stages for Correlation of Magallanes Basin Sediments

M. L. NATLAND
Consultant for ENAP
333 Weymouth Place
Laguna Beach, California 92651

EDUARDO GONZALEZ P.
Geologist for ENAP
Punta Arenas, Chile

ANTONIO CAÑON
ENAP Chief Micropaleontologist
Punta Arenas, Chile

MARIO ERNST
ENAP Assistant Micropaleontologist
Punta Arenas, Chile

1974

Copyright 1974 by The Geological Society of America, Inc.
Library of Congress Catalog Card Number 74-75964
I.S.B.N.. 0-8137-1139-8

Published by
THE GEOLOGICAL SOCIETY OF AMERICA, INC.
3300 Penrose Place
Boulder, Colorado 80301

*Printed in The United States of America
by Edwards Brothers, Inc., Ann Arbor, Michigan 48104.*

*The printing of this volume has been made possible
through the bequest of Richard Alexander Fullerton Penrose, Jr.,
and the generous support of all contributors
to the publication program.*

Contents

Part I
Geology and
Paleontology of Magallanes Basin
M. L. Natland and Eduardo Gonzalez P.

Abstract	3
Introduction	5
Previous work	7
General geology of the Magallanes Basin	9
Geomorphology, structure, and stratigraphy of the Magallanes Basin	11
System of stages	19
Correlation with other areas	53
Paleoecologic-sedimentary summary and conclusions	55
Acknowledgments	57

Part II
Magallanes Basin Foraminifera
Antonio Cañon and Mario Ernst

Abstract	61
Previous work	63
Magallanes Basin microfauna	65
Systematic micropaleontology	67
Acknowledgments	93
Plate section	95
Selected bibliography	109
Index	119

Figure
1. Geologic map of Magallanes province ⎫ .... inside back cover
2. Main morphostructural units in Magallanes Basin ⎭
3. After reconstitution of the Neocomian by R. Maack 14
4. Mazian Stage 22
5. Divisaderian Stage 23
6. MacPhearsonian Stage 24

7. Sebastinian Stage	25
8. Gaviotian Stage	26
9. Miradorian Stage	27
10. Rosarian Stage	30
11. Cameronian Stage	31
12. Moritzian Stage	32
13. Clarencian Stage	33
14. Brunswickian Stage	34
15. Manzanian Stage	35
16. Oazian Stage	36
17. Germanian Stage	39
18. Riescoian Stage	40
19. Lazian Stage	41
20. Peninsulian Stage	42
21. Tenerifian Stage	44
22. Pratian Stage	45
23. Esperanzian Stage	47
24. Rinconian Stage	49
25. Tobifera Series	50
26. Upper Paleozoic rock outcrops	51
27. Lower Paleozoic Series (crystalline basement)	52

Table .... inside back cover
1. Magallanes Basin biostratigraphic section
2. Magallanes Basin foraminiferal zones and some equivalents along the American continents and overseas
3. North-northwest–south-southeast cross section showing stage correlations in the Magallanes Basin
4. West-northwest–east-southeast cross section showing stage correlations in the Magallanes Basin
5. Distribution of Foraminifera and stages in the ENAP Sombrero no. 1 and Vania no. 1 wells. Ranges of key fossils occurring in other wells

Part I
Geology and Paleontology of Magallanes Basin

Abstract

Foraminiferal studies and paleoecologic analysis of the southern Patagonia (Chile) sedimentary sequence suggest that a system of stages, based on a time-stratigraphic consideration, can be used for correlation of Magallanes basin sediments. On the basis of Foraminifera and other characteristics, the following stages have been established for the Upper Jurassic-Cretaceous-Tertiary Magallanes sequence: Rinconian (Oxfordian-Kimmeridgian), Esperanzian (Portlandian-Hauterivian), Pratian (Barremian), Tenerifian (Aptian-Albian), Peninsulian (Albian-Cenomanian), Lazian (Cenomanian-Santonian), Riescoian (Santonian-Maestrichtian), Germanian (Paleocene-Danian), Oazian (Paleocene), Manzanian (lower-middle Eocene), Brunswickian (lower-middle Eocene), Clarencian (upper Eocene), Moritzian (upper Eocene), Cameronian (upper Eocene-lower Oligocene), Rosarian (upper Oligocene-lower Miocene), Miradorian (upper Oligocene-lower Miocene), Gaviotian (Miocene), Sebastinian (Miocene), MacPhearsonian (Miocene), Divisaderian (upper Miocene-lower Pliocene) and Mazian (Pleistocene). Lateral facies variations have been recognized in the Miradorian and Gaviotian stages, showing water shallowing from the southeast in Tierra del Fuego to the northwest in the mainland. It is probable that similar conditions exist in the lower Tertiary stages in the northern part of southern Argentina.

Introduction

This study is in two parts. Part I, by M. L. Natland and Eduardo Gonzalez P., presents a stage system of correlation for Magallanes Basin sediments and summarizes the general geology, biostratigraphy, and paleoecology of this basin. Part II, by Antonio Cañon and Mario Ernst, contains a systematic description of the microfaunal species, which has proved most useful for correlation, together with a review of previous paleontologic work.

In the past, stratigraphic units in the Magallanes Basin were established on the basis of either similar lithologic facies or similar faunal assemblages without regard for the fact that diverse units may have been deposited on the same time horizon. Consequently, many formations that are actually equivalent in time were not properly correlated because their time-stratigraphic relations were not known prior to this study. This same practice was followed during the early exploration of the Tertiary basins in California and led to inaccurate correlations resulting in oil exploration failures. Finally, a time-stratigraphic stage system was adopted in California which made more accurate correlations possible.

The stage system and conclusions in this paper were developed only after seven years of intensive study. More than 42,000 samples, taken at 3-m intervals from more than 60 closely spaced wells and key outcrop sections in the Magallanes Basin, were examined microscopically. Most of these samples are abundant in microfauna, particularly benthonic Foraminifera. Relative abundances were charted, not only of all foraminiferal species, but also of other constituents that might serve as zone markers or indicators of paleoecologic conditions.

A time-stratigraphic stage includes all sediments deposited during the same recognizable time interval. Since sedimentary processes vary widely, lithology of sediments on the same time horizon may change radically with distance. The same is true of marine faunas, all of which have preferred ecologic niches. As water deepens, with corresponding changes in temperature, salinity, and pressure, species also change. Therefore, for a stage system to be reliable, control wells must be spaced closely enough so that it may be reasonably concluded that lateral changes observed in the samples are caused by environmental or sedimentary variations within the same time unit.

Fortunately, in the Magallanes Basin, Empresa Nacional del Petróleo has drilled exploratory wells sufficiently close together to provide the control necessary to formulate a system of stages. Most of the samples from these wells are abundant in microfauna. The Tertiary microfauna has many modern analogs (living representatives) in the present oceans so that much is known of their ecologic requirements. It is possible, therefore, to determine approximately the conditions under which faunas in the following stages were deposited. In addition, the studies of modern planktonic Foraminifera, carried out by many paleontologists, offer much toward the evaluation and interpretation of Cretaceous assemblages in different basins. The Divisaderian Stage, however, has no fauna except mollusk fragments and diatoms in the eastern area, but it is characterized by abundant volcanic glass, which is an excellent time-stratigraphic indicator, because it was deposited from a widespread volcanic dust fallout.

The Sebastinian Stage has no identifiable fauna in the subsurface, except for occasional mollusk fragments, but it is characterized by abundant carbonaceous matter and concretionary phosphatic pellets which are fairly good time-stratigraphic indicators in the northeast part of the Magallanes province.

No serious attempt has been made at this time to correlate conclusively the Magallanes Basin faunas with standard European age assignments made for other basins to the north. The age assignments suggested in this report are tentative and subject to change.

Previous Work

In spite of rough terrain, remoteness, and the lack of good maps, the Magallanes area has been the subject of geological and paleontological investigations several times. Scientific expeditions from various nations have visited different portions of the province.

Until about 1925, most paleontological studies of southern South America sediments were restricted to megafossils. Based on early geological investigations, broad stratigraphic subdivisions were formulated by Phillippi (1887, 1896, 1898); Ortman (1898, 1899, 1900, 1902); Wilckens (1904, 1905, 1907, 1921, 1924); Paulcke (1906); Favre (1908); Stolley (1912); Hauthal and others (1907); Bonarelli (1917); and Felsch (1913, 1916) (Fig. 1).

Growing interest in finding new petroleum sources spurred more intensive geologic investigations by oil companies and government agencies. From 1920 to 1930, the Standard Oil Company, Royal Dutch Shell, and Pan American Company were active. After 1930 several Chilean government agencies, including Superintendencia del Salitre y Minas, Departmenta de Minas y Petróleo (DEMIPE), Corporación de Fomento (CORFO), and Empresa Nacional del Petróleo (ENAP), began intensive explorations.

After 1928 stratigraphic studies increased. The following authors have published works based on geological research for DEMIPE, CORFO, and ENAP: Keidel and Hemmer (1931); Hemmer (1936, 1937); Decat and Pomeyrol (1931); Ruby (1944); Thomas (1949a, 1949b); Wenzel (1951); Mordojovich (1951); Todd and Kniker (1952); Grossling (1953a); Cecioni (1955a, 1955b, 1955c, 1955d, 1956a, 1956b, 1957a, 1957b, 1959); Hoffstetter and others (1957); Fuenzalida (1942, 1964); Duhart (1963); Katz (1961a, 1961c, 1962, 1963, 1964); Katz and Watters (1965); Gutierrez (1962); Cortes (1964); Gonzalez and others (1965); Martinez (1968); and Cañon (1968). These publications provide excellent summaries of the stratigraphy and the main tectonic features of this region. In addition, many unpublished reports by geologists of Departmenta de Minas y Petróleo, Standard Oil Company, Corporación de Fomento de la Producción, United Geophysical Company, United Engineering Corporation, and Empresa Nacional del Petróleo are available in the files of the Chilean government petroleum agency.

General Geology of the Magallanes Basin

The Magallanes province may be divided into four physiographic regions from west to east (Gutierrez, 1962): (1) the Patagonian Archipelago, including the Pacific islands; (2) the Patagonian Andes (Main Cordillera); (3) the Andean foothill belt, which constitutes the large highland east and north of the Main Cordillera and forms several lower parallel ridges (the eastern end of the foothill belt closely coincides with the Tertiary-Cretaceous boundary); and (4) the pampas, or Magallanian steppe, extending east and north of the foothill belt toward the Atlantic.

South of the Gulf of Penas (47° S.) five morphostructural units can be distinguished along southern Patagonia (Harrington, 1965; Kranck, 1932; Ugarte, 1966; Fig. 2):

1. The Archipelago Mountain Range, which forms the row of islands along the Pacific Coast as far south as Cape Horn. It may represent the southern continuation of the Cordillera de la Costa, which is well developed in central Chile, Chiloe Island, Guaytecas, and the Chonos Archipelagos. This unit consists of dioritic rocks of the Andean suite, Paleozoic sedimentary rocks, and in some areas micaceous schists. Ophiolitic and basaltic rocks are also present on Hoste Island in the southernmost part of the archipelago (Cecioni, 1955b, 1955c; Cespedes, 1963, 1964, 1968; Cortes, 1968; Halpern, 1967; Katz and Watters, 1965; Kranck, 1932; Munoz Cristi, 1956).

2. The Main or Patagonian Cordillera, which extends as an unbroken mountain range immediately to the east of the Archipelago Mountain Range and then south nearly to the Strait of Magellan. From here the north-south trend changes to northwest-southeast, and the continuous mountain chain becomes a broken range penetrated by the sea in several places. It proceeds along Santa Ines Island, Clarence Island, Capitan Aracena Island, and the Darwin Cordillera. In the southwest part of Tierra del Fuego the mountains rise considerably, reaching the highest point (2,469 m) at Mount Luis de Saboya. The range then continues almost due west-east to Staten Island, located at the extreme end of South America. The core of the Main Cordillera is largely tectonized rocks. Some rock types, such as Jurassic volcanic rocks and patches of Cretaceous sediments, are intruded by ophiolites, granites, and dioritic rocks of the Andean suite and by Pliocene-Holocene effusive masses (Fig. 1).

3. The Magallanes Basin, which is described below in detail.

4. The Rio Chico Arch (Fig. 2) in northeastern Santa Cruz province, Argentina, which forms an elongate structural unit extending from the mouth of the Santa Cruz River in the southeast to Lake Buenos Aires in the northwest. The Rio Chico Arch has been a positive area since lower Paleozoic time. Rocks of this age crop out near La Modesta (Fig. 2) (Di Persia, 1960; Barwick, 1955; Harrington, 1965; Suero, 1962; Ugarte, 1966).

5. The upper Paleozoic-Jurassic basin, which extends to the northeast of the Rio Chico Arch into northern Santa Cruz and Chubut provinces, Argentina, and contains exposed igneous and continental rocks. Exposures include Permian granites at La Leona; Permian and Lower Cretaceous layers with microflora described by Archangelsky (1960a, 1960b) at La Juanita (Fig. 26), La Golondrina (Fig. 2), Estancia Leonardo and Piedra Sholte; Triassic continental beds at El Tranquilo (Fig. 2); Liassic continental sediments containing an *Otozamites* flora and an *Estheria* fauna interbedded with volcanics near Roca Blanca and First of April localities; Middle and Upper Jurassic volcanic rhyolitic rocks and tuffs interbedded with lacustrine sediments containing a *Cladophlebis, Equisetites* flora and an *Estheria, Notobatrachus* fauna at La Matilde and Chon Aike (Feruglio, 1949-1950; Harrington, 1965; Stipanicic and Reig, 1955; Suero, 1948, 1953, 1958, 1962; Suero and Criado, 1955; and Ugarte, 1966).

Geomorphology, Structure, and Stratigraphy of the Magallanes Basin

The Magallanes Basin is a prominent, negative, asymmetrical structural feature, lying between the Rio Chico Arch and the Main Cordillera (Figs. 1, 2). As known today, it extends over most of southern Patagonia (47° to 55° S.) and possibly as far north as Lake Nahuelhuapi (Fuenzalida, 1964). It covers half of the Magallanes province in Chile and part of the Santa Cruz and Tierra del Fuego provinces in Argentina. The basin is bounded on the west and south by the Patagonian Andes, on the east by a line about 150 km off the Atlantic Coast, and on the north by the Rio Chico Arch.

The basin section (Table 1) begins with the Tobifera Series, a continental Jurassic volcanic sequence, followed by a thick marine Upper Jurassic-Miocene sedimentary section overlain by continental Miocene-Pliocene tuff and sandstone beds with some thin, brackish water intercalations. These rocks are capped by Quaternary glacial and fluvioglacial deposits. The sequences rest on a Paleozoic basement complex and reach a maximum thickness of more than 9,000 m in northern Brunswick Peninsula.

The Springhill Platform (Fig. 2) lies in the northeast part of the Magallanes Basin, extending across the northeast corner of the Magallanes province in Chile into the Santa Cruz and Tierra del Fuego provinces in Argentina. It was a relatively stable shelf, dipping gently southwest, with slow deposition during Upper Jurassic-Cretaceous time. According to well and seismic data, the present Springhill Platform gradually ends at a hinge line that coincides approximately with the 4,000-m isobatic contour in Figure 2.

The mobile belt extends along the western part of the Magallanes Basin. This elongate, depressed area was structurally deformed and subsided rapidly to create a trough that was filled with sediments during Upper Jurassic-Tertiary time. These sediments were folded to form the Main Cordillera (Fig. 2).

Three structural belts may be distinguished in outcrops across the Magallanes Basin.

1. The northeast belt, where the dip varies from 0° to 10° in a predominantly northeast direction with some reversals. This attitude is observed mainly in the upper Tertiary outcrop belt in northern Tierra del Fuego and in the mainland north of Brunswick Península.

2. The central fold and thrust belt, which is characterized by sharp anticlines whose flanks range from vertical to low dips. This zone is confined to the area with thick Upper Cretaceous and lower Tertiary deposits.

3. The west and southwest belt, which is marked by overturned and recumbent beds and fairly low-angle thrust faults. This zone is confined to Upper Jurassic and Cretaceous deposits, and the contact of sedimentary strata with igneous and metamorphic rocks is considered to be its west and southwest boundary.

A core sample of basement rock from the María Emilia no. 2 well (Fig. 27) in northeast Tierra del Fuego has been isotopically dated by Halpern (1967, p. 3) at 267 ± 3 m.y. with a Sr^{87}/Sr^{86} initial ratio of 0.710. This result suggests that the granodiorite gneiss was formed during the upper Paleozoic but that the igneous rocks initially crystallized in Precambrian or lower Paleozoic time. The alteration and signs of shearing, shown by the basement rocks, suggest that the entire region was a positive area subjected to considerable weathering before being covered by Jurassic volcanic rocks. Granodiorite gneiss was also found in the Posesión no. 1, the Dungeness no. 1, and the Cormoran no. 1 wells (Fig. 27).

In lower Pennsylvanian time the sea invaded the Patagonian Archipelago area, which is composed of schist and metamorphic rocks of possible lower Paleozoic age. As a result, a thick section of Pennsylvanian–Lower Permian fusilinid-bearing limestones, intercalated with shales and sandstones, was deposited (Cecioni, 1955b, 1955c; Hollingsworth, 1954). These sediments extend along the outer row of islands between the Gulf of Penas and the Strait of Magellan to form the Madre de Dios Basin (Fig. 2). Farther northeast in western Chubut, Argentina, the Pennsylvanian is represented by the marine Tepuel group (Fig. 26; Suero, 1958). As in most of southern South America, the Upper Permian was essentially a positive epoch in Patagonia (Harrington, 1962).

During the Triassic, southern Patagonia remained emergent (Harrington, 1962; Munoz Cristi, 1956). In northern Santa Cruz province, Argentina, the Triassic is represented by conglomerates, sandstones, shales, and a basic volcanic continental sequence exposed near El Tranquilo (Fig. 2; Ugarte, 1966). Along the east and northeast slope of the Main Cordillera the Triassic may be represented by thick, sandy conglomerates, tuffs, and a quartzite continental sequence exposed in the Monte Tres Picos area on the south coast of Brunswick Península (Cespedes, 1965), in Ultima Esperanza, and in southern Tierra del Fuego. These questionable Triassic rocks overlie the schists of the Main Cordillera and may underlie the Jurassic volcanics.

The Lower and Middle Jurassic was essentially a geocratic period in southern Patagonia south of the Rio Chico Arch. It is marked by rhyolitic flows, ignimbrites, and tuffs indicative of volcanism on an unprecedented scale. This rock sequence, called the Tobifera Series, is mainly volcanic rocks with inliers of Paleozoic granodiorite gneiss and schist. Tobifera relief is pronounced in the ignimbrites of the Cullen-Tres Lagos area (Fig. 2) in northeast Tierra del Fuego but gentle in the pyroclastic beds of the Peninsula Espora and Punta Delgada areas. Tobifera volcanic rocks (Fig. 25) are associated with water-laid continental beds and cover most of the Magallanes Basin.

On the basis of Foraminiferal data (Sigal, 1967), the Magallanes Basin was formed by a negative differential movement that caused instability of the earth's crust during the Upper Jurassic and Oxfordian. On the basis of megafossils the basin was formed during Portlandian (Tithonian) time (Cecioni, 1951, 1955a; Collignon, 1957; Cortes and Cañon, 1964; Feruglio, 1949-1950; Grossling, 1953a; Harrington, 1962). This downwarping permitted the sea to invade Patagonia through a trough opening southward (Munoz Cristi, 1956). During Oxfordian-Kimmeridgian time the Magallanes Basin was a pericratonic basin (Harrington, 1962) with a wide, stable shelf extending over the present extra-Andean Patagonia and a narrow, unstable shelf (Cortes, 1964, 1969) in the Ultima Esperanza area (Fig. 2). The narrow shelf received sediments mainly from a western source, probably located in the present Main Cordillera and Archipelago Range. This feature may be the miogeanticlinal ridge of the Upper Jurassic basin located along the eastern slope of the Main Cordillera. The western boundary of this primitive basin is unknown but is probably west of the Main Cordillera.

Magallanes Basin fill (Table I) begins with sedimentary rocks of the Rinconian Stage (Oxfordian-Kimmeridgian), which were deposited mainly on Tobifera volcanic rocks, although in the Sena Poca Esperanza area at Ultima Esperanza they rest on schists of the Main Cordillera. The Rinconian sequence, composed of light-gray sandstone, silty shale, and siltstone, is extensively developed (Fig. 24) along the east and north slope of the Main Cordillera and on the Springhill Platform, where the basal sediments are fluviatile deposits filling the depressions of the Tobifera paleorelief (Harrington, 1965; Montadert, 1968).

Sigal and others (1970) have pointed out the similarities of Upper Jurassic-Lower Cretaceous (Neocomian) microfaunas in Africa and Chile. Such distinctive and restricted species as *Reinholdella* cf. *quadrilocula* Subbottina and Datta and *Astacolus microdictyotos* Espitalie and Sigal are common to the Upper Jurassic of both Madagascar and the Magallanes Basin, where they occur in the Springhill formation (Rinconian). This fact suggested to Sigal and others (1970) that the Magallanes Basin may have been a southern extension of a relatively narrow trough separating the South America-Africa continental mass from the Antarctica-India-Australia continental mass prior to more extended separation (Fig. 3).

Marine sedimentation continued in the Magallanes Basin with euxinic conditions prevailing during the Esperanzian and Pratian Stages (Portlandian-Barremian). As a result, dark brownish-gray laminated shale beds of these two stages are widespread (Figs. 22, 23) (Barwick, 1955; Cecioni, 1955a, 1955b; Cortes, 1960; Fuenzalida, 1964; Grossling, 1953a, 1953b; Harrington, 1962; Katz, 1963; Scott, 1964; Ugarte, 1966).

During the Tithonian-Neocomian epoch, the eugeosynclinal trench extended over the present-day Main Cordillera area and the sedimentary belt to the west and south of this mountain range (Katz, 1964). The primitive stable shelf ended gradually at a hinge line that coincided approximately with the eastern and northern part of the present-day foothill belt (to the south of Evans no. 1 well). To the east and north of the sub-Andean zone the miogeosynclinal realm of the Magallanes geosyncline developed, while along the Main Cordillera and the areas to the west and south of this range the eugeosynclinal realm of the Magallanes area geosyncline developed (Katz, 1964).

During Lower Cretaceous time, a basic igneous activity began in the eugeosynclinal trench along the present-day Main Cordillera and the Hoste-Navarine Island (Cespedes, 1960; Katz, 1964; Katz and Watters, 1965). According to Kranck (1932) these activities occurred contemporaneously with earlier stages of folding in the Main Cordillera. Katz and Watters suggest that these rocks were emplaced simulta-

Figure 3. After reconstitution of the Neocomian by R. Maack.

neously with the Yahgan sediments (Lower Cretaceous). This simatic igneous phase, which ended during Upper Cretaceous time (Cespedes, 1963), was marked by submarine ophiolitic emissions of altered basic and intermediate igneous rocks (mainly sills) which were emplaced in the Lower Cretaceous beds (Cespedes, 1968; Katz and Watters, 1965; Kranck, 1932). An isotopic age measurement by Halpern (1967), of an ophiolitic rock sample from Bertrand Island, gave a K-Ar date of 92.5+ m.y.

Marine sedimentation continued without interruption during the Peninsulian-Pratian stages (Figs. 20, 22). The marine invasion reached its maximum expansion during the Peninsulian and Tenerifian stages (Figs. 20, 21). These sediments are light gray and reddish limy shales with some chalky shale containing occasional algal deposits and light-gray shale with white specks. They are extensively developed over Santa Cruz and Magallanes provinces (including Tierra del Fuego).

During the Lazian Stage (Campanian-Cenomanian) the marine sedimentation continued in the Magallanes province and in the greater part of Santa Cruz province. As a result of intra-Lazian-Peninsulian movements, the sub-Hercynian orogeny postulated by Cecioni (1959) affected mainly the belt west of the present-day Main Cordillera. As a consequence, part of the Patagonian Archipelago area was uplifted (Katz, 1961c, 1962, 1964) and the trough of the Upper Cretaceous basin subsided rapidly. The axis of the Lower Cretaceous basin migrated to the east and northeast. The stable shelf developed a strong regional tilt toward the west and south from the western boundary of the present-day Springhill Platform. Along the Main Cordillera, a great thickness of gray, greenish-gray, and hard, dark shales of the Lazian Stage (Fig. 19) were deposited. In the Ultima Esperanza area and farther south along the foothill belt, the Lazian Stage is characterized by a sequence of turbidites with thick lenticular interbeds of conglomerate (gravitites) (Natland, 1967) with a mudstone matrix, deposited by subsea gravity flows (for example, Sofia conglomerate, in Ultima Esperanza; Cerro Diadema conglomerate, in Skyring Sound; Valdes conglomerate, in Dawson Island; and Cerro Flecha and the Cerro Colo-Colo conglomerates, in Tierra del Fuego). Sedimentary structure indicates a general eastward paleoslope. South- and southeast-flowing paleocurrents show that the main source that shed sediments eastward onto the Patagonian block was the Andean Island Arc system (Cortes, 1960; Scott, 1966).

The emplacement of the Andean intrusive suite (largely diorite, granodiorite, monzonite, and tonalite, with some acid and basic derivatives) in the Archipelago area and in parts of the Main Cordillera began during the Lazian, in the postorogenic period (Cespedes, 1963, 1964; Halpern 1962, 1967; Kranck, 1932; Katz, 1964; Katz and Watters, 1965). This sialo-simatic magmatic suite was confined exclusively to the internal zones (eugeosynclinal realm). Isotopic age measurement for rocks of the Andean intrusives by Halpern (1962, 1967) gave Rb-Sr isochron dates of 77 ± 2 m.y. to 82 ± 3 m.y. Marine accumulation continued uninterruptedly in the Magallanes and southern Santa Cruz provinces until the end of the Cretaceous, with the deposition of dark-gray, greenish shale, siltstone, sandy shale, and sandstone of the Riescoian Stage (Fig. 18).

In the late Riescoian time, a general regression occurred in the northern part of the Magallanes province with deposits of a thick sequence of molasse type sediments (Cecioni, 1957a). The sedimentary structures, lithologic types, fauna and flora indicate over-all shallowing within the Ultima Esperanza area and the filling of the trough from a west to northwest source (Cortes, 1964; Scott, 1905). Southward on Brunswick Península and Dawson Island, along the foothill belt, the marine flysch type accumulation persisted until Riescoian time (Rocallosa formation).

Lithologically, it is difficult to establish the Cretaceous-Tertiary boundary in the Magallanes Basin because of the apparent discontinuous sedimentation observed in outcrop along the foothill belt and in exploratory wells. Also adding to the difficulty is the uncertainty regarding age of some formations considered as either late Riescoian or Oazian. During this interval, a sequence of glauconitic sandstone and hard, dark-greenish shale of the Germanian Stage was deposited (Fig. 17).

Sometime between late Riescoian and Brunswickian, probably in the Germanian, tectonic movements affected the western mobile belt of the basin—the phase of

Laramian orogeny postulated by Katz (1961a, 1962). These Laramian orogenic movements were confined mainly to the present-day mountain range along the western and southwestern belt of the Magallanes Basin. As a result, the Patagonian Cordillera was uplifted to a positive position for the first time. Because of this activity, the Rinconian-Riescoian sequence was affected by secondary folding and faulting. After these events, the landmass contributing sediments to the Magallanes Basin was a newborn Cordillera that shed sediments eastward and northward into the lower Tertiary trough (Katz, 1961c). At this time the Magallanes Basin, as we know it today, came into existence. These Laramian orogenic movements were intense in southern Tierra del Fuego, where a thick deposit of intra-Eocene conglomerates (Ballena formation) overlies early Eocene (Manzanian) shales (Barwick, 1955). In the Paleocene (Oazian), probably in the postorogenic period, the Darwin Cordillera intrusives were emplaced.

Sedimentary structures and thickness data suggest that the axis of the successively younger troughs in the Cretaceous and Tertiary migrated progressivly farther east and northeast with time.

In the lower Tertiary (Germanian-Cameronian), marine deposits were restricted to the foredeep east and northeast of the rising Main Cordillera. Persistent sedimentation through Oazian to Rosarian Stages, in most of the Magallanes and southern Santa Cruz provinces, deposited a thick sequence (more than 3,800 m in ENAP Manzano no. 7 well) composed of sediments ranging from claystones to conglomerates of various colors and textures (Figs. 10 to 16).

In northern Ultima Esperanza and northern Santa Cruz province, the lower tertiary beds are mostly nonmarine or brackish water (Cortes, 1964; Feruglio, 1949-1950). During the Clarencian Stage (Fig. 13), a partial regression took place in the central part of Riesco Island and in southern Ultima Esperanza. During this regression, the coaly beds and cross-bedded deltaic sandstones of the Tres Brazos formation were deposited.

At the close of the Miradorian through Cameronian, a regression took place in the western part of the Magallanes Basin, and the sediments of the Rosarian Stage were overlain by continental coal-bearing beds and lagoonal, beach, and open marine deposits of the Miradorian Stage (Fig. 9).

During the upper Tertiary (Miocene-Pliocene) the Magallanes Basin was further reduced by the progressive rising of the Patagonian Cordillera. Marine conditions remained in the eastern part of Magallanes province, mainly in eastern Tierra del Fuego, and persisted until middle Miocene with the deposits of light-brown claystone, silty claystone, shaly sandstone, and conglomerate of the Gaviotian Stage (Fig. 8). The subsurface data suggest that the bathymetric axis of the basin's trough extended, during this stage, from northeast to southwest through the eastern part of the Magallanes Basin (Cañon, 1968). In late Miocene, the Gaviotian trough was filled with continental beds and brackish water, and thus was obliterated. The Magallanes Basin area was raised above sea level, and its central and eastern parts accumulated deposits of sandstone, conglomeratic sandstone, and claystone with lignitic beds of the Sebastinian Stage (Fig. 7).

In late Miocene, the sea invaded a narrow strip along the Atlantic coast of southern Patagonia and deposited fine-grained siltstone and claystone bearing shell fragments of the MacPhearsonian Stage (Fig. 6). Westward, the MacPhearsonian strata are largely of nonmarine or brackish water origin.

A last major diastrophic phase took place at the close of the MacPhearsonian Stage in the Magallanes province, and was responsible for the upwarping of the foothill belt in southern Tierra del Fuego, Dawson Island, Brunswick Península, Riesco Island, and the southern Ultima Esperanza areas. Subsidiary folding also occurred. Toward the east a rather gentle folding cycle began. The movements in the foothill belt and in Ultima Esperanza were accompanied by the intrusion of Cerro Paine and Cerro Donoso plutonic bodies (Katz, 1961a, 1962; Quensel, 1913). The isotopic age measurement (Rb-Sr) of Cerro Paine adamellite given by Halpern (1967) is 12 ± 2 m.y. Farther south the igneous activity is represented by the alkali-rich basic effusive rocks, mainly trachydolerite (Quensel, 1913) of the Cordillera Pinto-Cordillera Vidal area to the north of Skyring Sound, and Cerro Fraile in Riesco Island. The Cordillera Pinto area is one of the main volcanic areas which produced the pyroclastic debris of the Divisaderian Stage (Fig. 5). Because of this volcanic activity, which occurred along the eastern edge of the foothill belt, deposits of tuffaceous beds bearing vertebrate remains, bluish sandstone, and some flows of basaltic lava of the Divisaderian Stage were spread eastward onto the extra-Andean Patagonia and on northern Tierra del Fuego (Gonzalez, 1953; Scott, 1903, 1905, 1928; Simpson, 1941; Sinclair, 1901-1906).

As a consequence of the final uplift of the main Cordillera, which occurred at the close of the Pliocene epoch, a final compression began in the mobile belt of the basin, and wrench faulting then affected the foothill belt (Katz, 1962). As a result of these predominantly vertical movements, a gentle folding cycle was then responsible for upwarping the Rinconian-Divisaderian stratigraphic sequence east of the foothill belt.

The volcanic activity continued in the Magallanes province during Pliocene-Pleistocene time, with flows of olivine basalt and showers of volcanic ash in scattered places along the foothill belt (Cerro Baquales) and occasionally in the Archipelago Range area between Navarino Island and Ultima Esperanza.

The deterioration of climatic conditions and the rising of the Patagonian Cordillera marked the coming of Pleistocene glaciations. This glacial activity then produced the final touches in the shaping of present-day Magallanes province landscape. Deposits of glacial drift of the Mazian Stage (Fig. 4) covered a great part of the Magallanes province (Brueggen, 1929; Caldenius, 1932). During this stage, deposits of glacial debris containing boulder clay, varved clay, sand and gravel, attained their maximum thickness and importance in the extra-Andean area. (A thickness of 232 m was recorded in the Rio del Oro no. 1 well.)

Volcanic activity was reactivated during the Holocene. It is represented by basaltic lava flows, ash beds, and volcanic cones, which overlie and protrude through the glacial deposits in the extra-Andean Patagonia located in the northeast portion of the Magallanes province, along the international boundary, and the southernmost part of Santa Cruz province (Fig. 1). Some active volcanos observed in the central Cordillera area belong to the same volcanic cycle. A general westward tilting affected southern Patagonia during Pleistocene-Holocene time, which caused some east-flowing rivers to change courses and flow westward into the Pacific drainage system.

System of Stages

The stages described below permit correlation of all sections, according to time, by careful consideration of the microfauna and lithologic characteristics.

QUATERNARY

Pleistocene

Mazian Stage
   Characteristics and Significant Fossils. This stage was deposited over the eastern part of the basin (Fig. 4). It is composed of fine-to-coarse glacial deposits cemented with clay, plus marine terrace sediments along eastern Brunswick Península. The marine beds contain a shallow-water molluscan and foraminiferal fauna in which *Elphidium crispum* Linné is persistently present.
   Paleoecology. Marine depth of 1-10 m.
   Type Surface Section. Juan Mazia Peninsula, Tierra del Fuego, and Cabo Negro in northeast Brunswick Península.
   Type Subsurface Section. Clarencia no. 1A, 0-225 m.

TERTIARY

Upper Miocene-Lower Pliocene

Divisaderian Stage
   Characteristics and Significant Fossils. This stage is limited to the northeastern portion of the basin (Fig. 5) and is thickest (400 m) in the Penitente River-Cordillera Vidal area. It is composed of tuff, volcanic ash, and volcanic glass interbedded with bluish-gray sandstone and conglomerate. There is no marine fauna except for some mollusk fragments and diatoms in the Peninsula Espora and Catalina-Dungeness areas. Characteristic fossils are: *Nematherium birdi* and *Astrapotherium magnum* (vertebrates), mollusk fragments, diatoms, silicified wood, *Nothofagus*, and volcanic glass.
   Paleoecology. Continental, except for very shallow (1-10 m) marine conditions in the areas mentioned above.

Type Surface Section. Palomares hills.
Type Subsurface Section. Punta del Cerro no. 1, 55–300 m.

Miocene

MacPhearsonian Stage

Characteristics and Significant Fossils. This stage also covers the northeast part of the basin (Fig. 6) and is thickest (550 m) in the El Salto no. 1 area. It is composed of coarse greenish-gray sandstone, conglomerate, and greenish-gray claystone. Diagnostic fossils are: *Buccella depressa* Andersen, *Nonionella auris* (d'Orbigny), *Nonionella pulchella* Hada, and *Trifarina angulosa* (Williamson) in association with mollusks, ostracods, echinoid spines, glauconite, and abundant carbonaceous material.

Paleoecology. Final marine transgression with depths of 1–30 m. Water was deepest in the southeast part of the stage area, shallowing to non-marine toward the northwest.

Type Surface Section. MacPhearson hill, Tierra del Fuego.
Type Subsurface Section. Pampa Larga no. 1A, 505–802 m.

Sebastinian Stage

Characteristics and Significant Fossils. This stage extends over the northeast part of the basin (Fig. 7) and is thickest (279 m) in the Pampa Larga no. 1A area. It is composed of greenish-gray and brown claystone interbedded with lignite and conglomerate. There are no Foraminifera, except for a few specimens that may be drilling contaminants from the MacPhearsonian or may be from very localized marine sediments. Characteristic fossils are: *Ostrea* d'Orbigny, diatoms, glauconite, mollusk fragments, abundant carbonaceous material, and concretionary phosphatic pellets.

Paleoecology. Mostly continental deposits with some shallow marine sediments in lagoons, beaches, and intertidal areas.

Type Surface Section. Estancia Filaret–Bandurria area, Tierra del Fuego.
Type Subsurface Section. Pampa Larga no. 1A, 802–1,081 m.

Gaviotian Stage

Characteristics and Significant Fossils. This stage covers the eastern part of the basin (Fig. 8) and is thickest (320 m) in the Gaviota Lake–Filaret area. It is composed of light-gray silty claystone, sandy at the base. As shown in Figure 8, there are two ecological provinces, west and east. The west province contains a lagoonal-shallow-water fauna which includes: *Ostrea* d'Orbigny, abundant mollusk fragments, carbonaceous material, *Buccella depressa* Andersen, *Florilus scaphus* (Fichtel and Moll), and *Florilus* cf. *boueanus* (d'Orbigny). The east province has a deep-water fauna including: *Praeglobobulimina pupoides* (d'Orbigny), *Hoeglundina elegans* (d'Orbigny), *Pullenia bulloides* (d'Orbigny), *Cassidulina* cf. *brocha* Poag, *Cyclammina cancellata* Brady, *Gyroidina soldanii* d'Orbigny, and *Robertina arctica* d'Orbigny. Abundant shallow-water megafossils in concretions are also found in the east province, but their occurrence with such deep-water species as *Pullenia bulloides* and *Cyclammina cancellata* indicates that the megafossils are not indigenous but were transported downslope from shallow to deep water.

Paleoecology. Penultimate marine transgression, ranging from shallow (1-30 m) in the west to deep (1,000 m) in the east. In general, the fauna from key sections indicates a gradual deepening of the marine environment from northwest to southeast. In the west province, the presence of megafossil fragments plus carbonaceous material suggests a lagoonal environment that gradually deepens toward the east, where species such as *Florilus scaphus* and *Buccella depressa*, found in the El Salto no. 1, the San Antonio no. 1, and the Manzano no. 7 wells, suggest water depths of 1-30 m. Because the southwestern extension of the west province was uplifted and eroded away, the geologic record was lost.

In the east province, the Sombrero no. 1 contains *Hoeglundina elegans* and *Cassidulina* cf. *brocha*, pointing to water at least 300 m deep. The Cisne no. 1 well, near Gaviota Lake, has a mixture of deep- and shallow-water faunas, but species such as *Pullenia bulloides* and *Cyclammina cancellata* indicate that the displaced shallow-water forms were transported downslope into water 1,000 m deep.

Type Surface Section. (a) West province: Estancia El Salto area northeast of Seno Skyring; (b) east province: Estancia Gaviota-Brush Lake area, central Tierra del Fuego.

Type Subsurface Section. (a) West province: Manzano no. 7, 873-1287 m; (b) east province: Filaret no. 1, 423-540 m.

Lower Miocene-Upper Oligocene

Miradorian Stage

Characteristics and Significant Fossils. This stage extends over the eastern part of the basin (Fig. 9) and is thickest (580 m) in the Carmen-Rio Chico area. It is composed of greenish-gray, fine-to-coarse glauconitic sandstone interbedded with light-gray siltstone and silty claystone. There are also two ecologic provinces that nearly coincide with those of the Gaviotian except in the Cruceros-Punta del Cerro area. Here the Gaviotian east province marine fauna is found in the Cruceros no. 1 well, but the Miradorian in this well contains only continental and lagoonal deposits of the west province. This fact indicates that the marine transgression progressed farther west during the Gaviotian than during the Miradorian.

Miradorian west province fossils are: mollusk fragments, abundant carbonaceous material, and rare arenaceous Foraminifera. The only identifiable foraminiferal species is a *Trochammina* probably related to *T.* cf. *inflata* (Montagu), the most common *Trochammina* found in modern lagoons. The Magellanian megafossil beds in the Minas River Valley-Mirado hills area west of Punta Arenas have been included in the Miradorian, and their fauna has been discussed by several authors (Feruglio, 1949-1950; Fuenzalida, 1942; Ortmann, 1899; Phillippi, 1887).

Miradorian east province fossils include: *Hoeglundina elegans* (d'Orbigny), *Sphaeroidina bulloides* d'Orbigny, *Pullenia bulloides* (d'Orbigny), *Gyroidina soldanii* d'Orbigny, *Karreriella cushmani* Finlay, and *Psamminopelta venezuelana* (Hedberg). In the deeper water area from the Sombrero no. 1 to Gaviota Lake, the top of the Miradorian is marked by *Psamminopelta venezuelana*.

Paleoecology. Marine transgression ranging from shallow (1-30 m) in the west to deep (1,000 m) in the east. Distribution of Miradorian fauna indicates a gradual deepening of the basin from west to east in the same manner as the Gaviotian

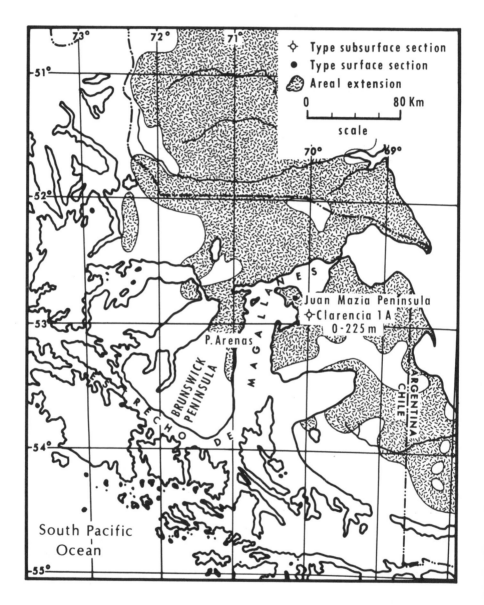

Figure 4. Mazian Stage

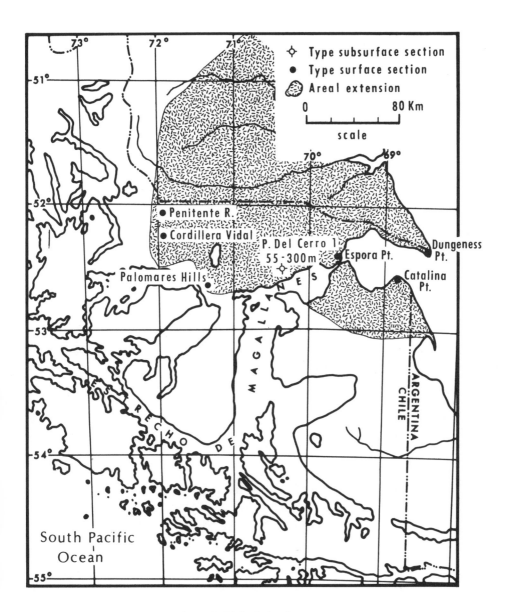

Figure 5. Divisaderian Stage

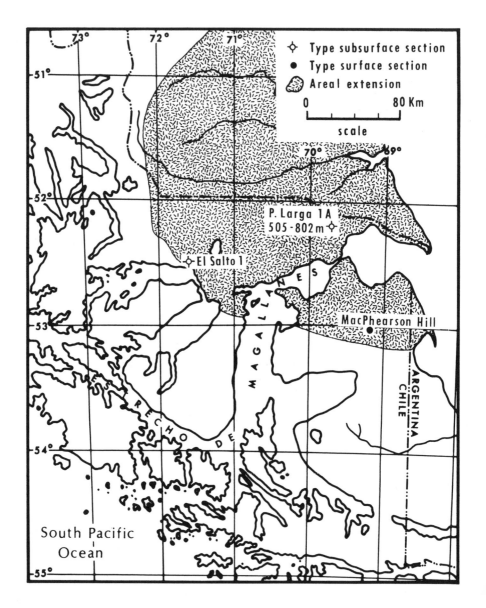

Figure 6. MacPhearsonian Stage

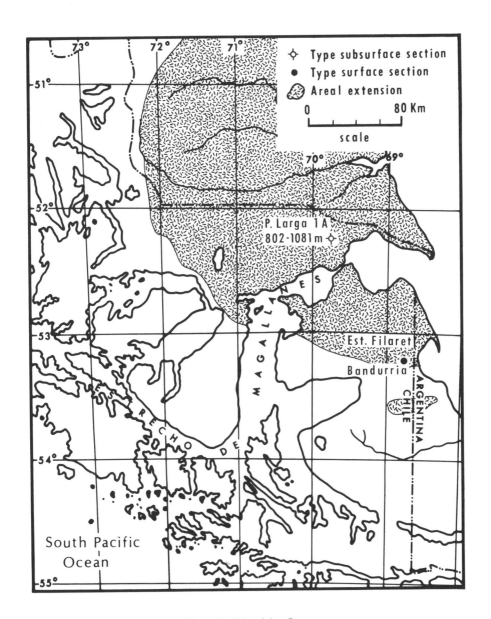

Figure 7. Sebastinian Stage

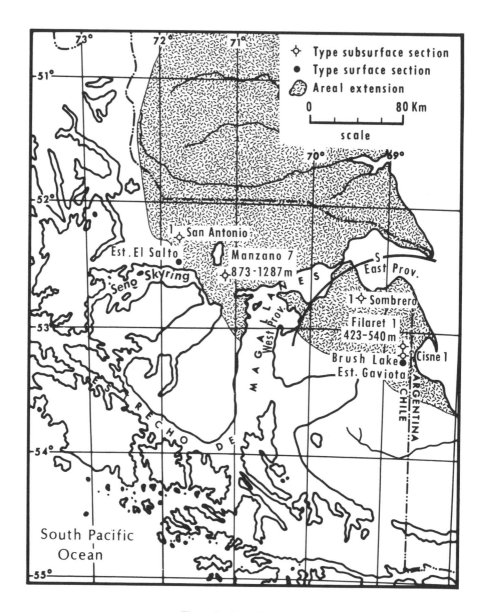

Figure 8. Gaviotian Stage

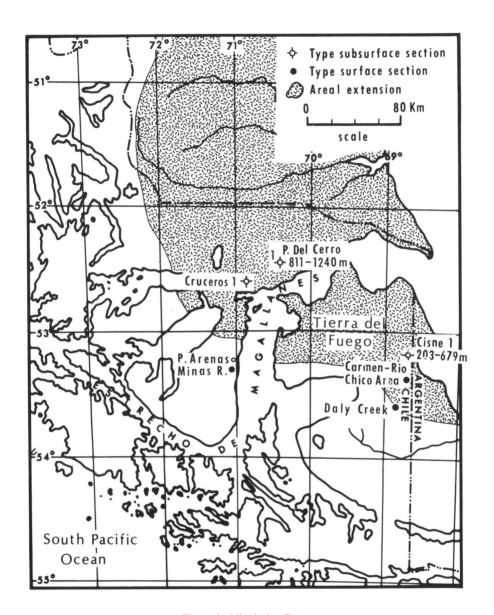

Figure 9. Miradorian Stage

except that, in the latter, the marine transgression progressed farther west. The west province fauna of the Miradorian suggests lagoonal conditions, particularly the presence of *Trochammina* cf. *inflata*. The depositional environment of the Miradorian in the east province is much like that of the Gaviotian, with shallow, open-sea conditions in the western margin and a rapid deepening toward the Gaviota Lake and Rio Chico areas. Such species as *Hoeglundina elegans, Sphaeroidina bulloides, Gyroidina soldanii, Karreriella cushmani,* and *Pullenia bulloides* indicate oceanic depths of 500–1,000 m.

Type Surface Section. (a) West province: Minas River Valley–Mirador Hills area; (b) east province: Daly Creek area near Rio Chico, Tierra del Fuego.

Type Subsurface Section. (a) West province: Punta del Cerro no. 1, 811–1,214 m; (b) east province: Cisne no. 1, 202–679 m.

Rosarian Stage

Characteristics and Significant Fossils. This stage covers much of the basin (Fig. 10) and is thickest (550 m) near the Manzano no. 5. It is composed of light-gray sandy siltstone and silty sandstone with abundant glauconite. Thick lenticular fine-grained sandstones are locally developed. Foraminifera are generally abundant in this stage, but their population is greatly reduced where there is a high percentage of sandstone. In the southeastern part of the basin, the stage top is marked by the first occurrence of *Rectuvigerina ongleyi* (Finlay) and the top of the second prolific zone of *Psamminopelta venezuelana* (Hedberg). In the western part of the basin, the stage top is marked by the first occurrence of a thin-walled arenaceous foraminiferal species assigned to *Gaudryina* in this paper. It is usually found badly crushed but with characteristics sufficiently distinct to permit consistent identification.

Paleoecology. Marine depth of 500 m with open-sea conditions prevailing in all wells studied.

Type Surface Section. Upper valley of Rosario Creek.

Type Subsurface Section. Punta del Cerro no. 1, 1,214–1,622 m.

Lower Oligocene–Upper Eocene

Cameronian Stage

Characteristics and Significant Fossils. This stage was deposited over much of the basin (Fig. 11) and is thickest (750 m) in the Cruceros area. It is composed predominantly of light-gray silty claystone and fine-grained silty sandstone with some locally developed thick lenticular fine to medium sandstones (Tables 3 and 4). The top of this stage is marked by the first occurrence of *Marginulina knikerae* Cañon and Ernst, n. sp., by the first persistently abundant occurrence of *Virgulinella severini* Cañon and Ernst, n. sp., and by *Pullenia bulloides* (d'Orbigny). Because the basin shallows toward the northwest, *Virgulinella severini* becomes progressively higher in the section and occurs sporadically in the Rosarian.

Paleoecology. Marine depth of 1,000–1,500 m.

Type Surface Section. Ciervos and Leña Dura River valleys in Brunswick Península and northern coast of Bahía Inútil near the mouth of the Discordia River in Tierra del Fuego.

Type Subsurface Section. Punta del Cerro no. 1, 1,611–2,102 m.

Upper Eocene

Moritzian Stage
Characteristics and Significant Fossils. This stage is not as extensive as the Cameronian and covers the eastern part of the basin (Fig. 12). It is thickest (more than 500 m) in the San Antonio no. 1 and El Salto no. 1 wells but thins rapidly toward the east and disappears before reaching the Pampa Larga no. 1A well (Tables 3 and 4). The Moritzian is composed mostly of light-gray claystone. The top of this stage is marked by the first occurrence of *Plectina elongata* Cushman and Bermúdez and by the first persistent, very abundant occurrence of Radiolaria, *Spumellaria*? sp. 1.
Paleoecology. Marine depth of 1,000–2,000 m. The increase in Radiolaria and planktonic forms, relative to benthonic Foraminifera, indicates water deeper than that of Cameronian time.
Type Surface Section. Leña Dura River valley in Brunswick Península.
Type Subsurface Section. Punta del Cerro no. 1, 2,102–2,393 m.

Clarencian Stage
Characteristics and Significant Fossils. This stage extends over most of the basin (Fig. 13) and is thickest in the Manzano area. Although, in general, mainly claystone and silty claystone, it is predominantly sandstone and claystone in the western part of the basin, conglomerate in the Ballena Hills in southern Tierra del Fuego, and cross-bedded sandstone in the Ultima Esperanza area. The top of this stage is marked by the first occurrence of *Tritaxia chileana* (Todd and Kniker), *Globigerina triloculinoides* Plummer, and *Elphidium patagonicum* Todd and Kniker. In addition, *Lenticulina* cf. *asperuliformis* (Nuttall) is persistently present and Radiolaria *Spumellaria*? sp. 5 becomes very abundant.
Paleoecology. Marine depth of 1,000–2,000 m, with conditions much the same as the Moritzian.
Type Surface Section. Tres Brazos River valley.
Type Subsurface Section. Punta del Cerro no. 1, 2,393–2,519 m.

Lower Eocene–Middle Eocene

Brunswickian Stage
Characteristics and Significant Fossils. This stage also covers most of the basin (Fig. 14) and is thickest (more than 800 m) in the Manzano area. It is composed almost entirely of dark-gray silty claystone. The top of this stage is marked by the first occurrence of *Spiroplectammina adamsi* Lalicker and *Dorothia principensis* Cushman and Bermúdez in association with *Hastigerina iota* (Finlay) and *Globorotalia* cf. *crassata* (Cushman) var. *aequa* Cushman and Renz.
Paleoecology. Marine depth of 1,000–2,000 m.
Type Surface Sections. Tres Brazos and Santa Maria River valleys.
Type Subsurface Section. Punta del Cerro no. 1, 2,519–2,837 m.

Manzanian Stage
Characteristics and Significant Fossils. This stage is limited to a north-south swath through the center of the basin (Fig. 15) and is thickest (more than 2,000 m)

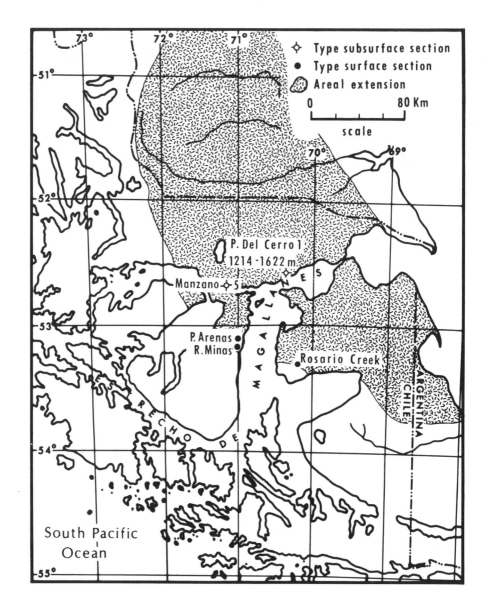

Figure 10. Rosarian Stage

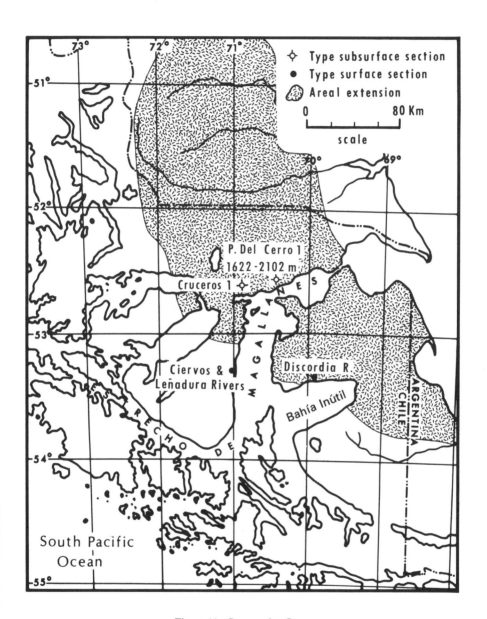

Figure 11. Cameronian Stage

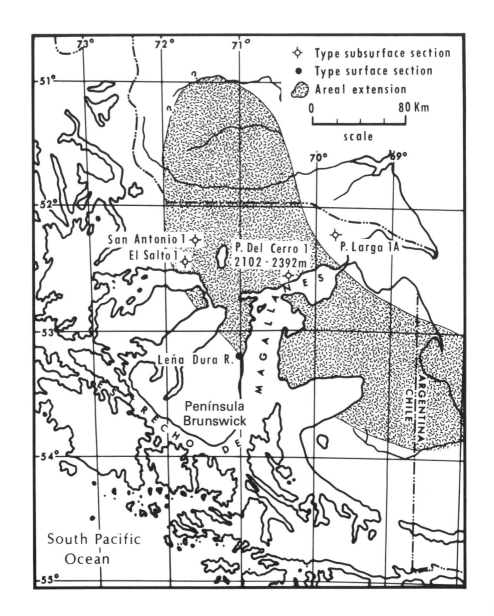

Figure 12. Moritzian Stage

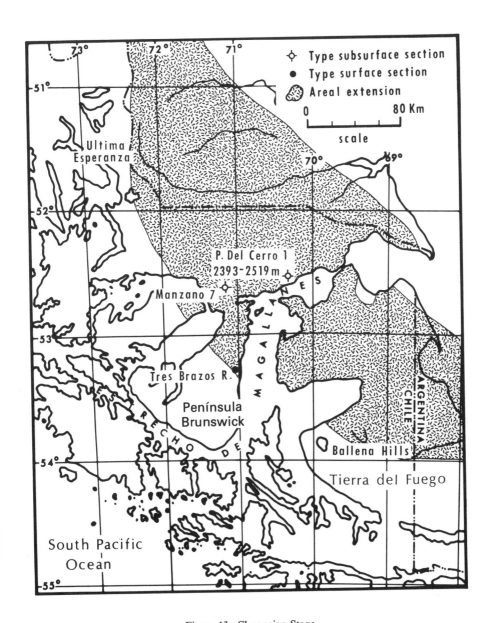

Figure 13. Clarencian Stage

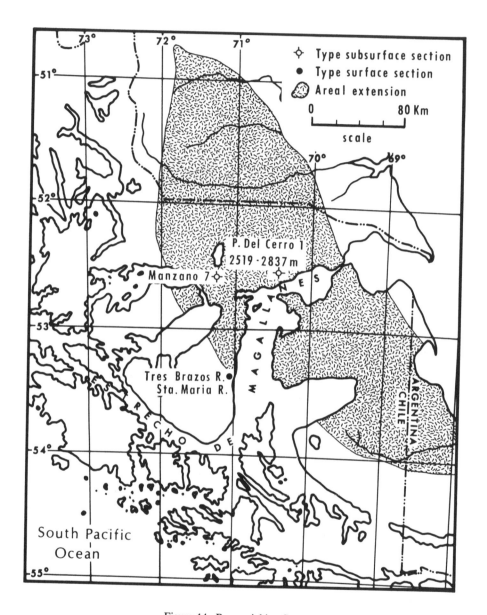

Figure 14. Brunswickian Stage

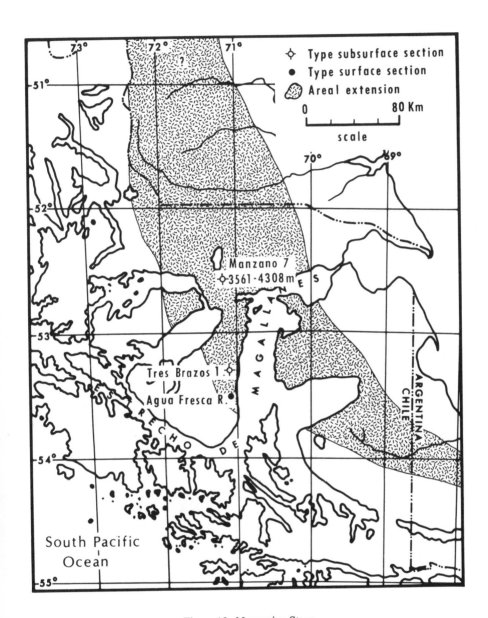

Figure 15. Manzanian Stage

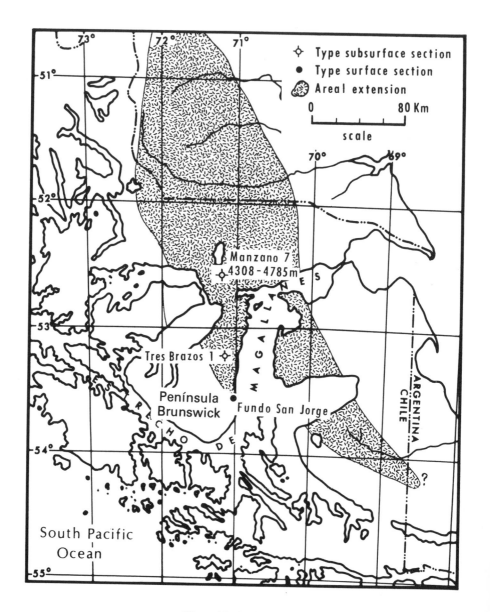

Figure 16. Oazian Stage

in the Tres Brazos no. 1 well. It is composed of massive gray claystone and siltstone with numerous large limestone concretions and some dense brown lime. Several sandstones, aggregating 50 m, occur in the central part of the basin. The top of this stage is marked by the first occurrence of *Candeina cecionii* Cañon and Ernst, n. sp., *Allomorphina conica* Cushman and Todd, and *Elphidium aguafrescaense* Todd and Kniker. *Spiroplectammina brunswickensis* Todd and Kniker also becomes conspicuously more abundant.

Paleoecology. Marine depth of 1,000–2,000 m.

Type Surface Section. Agua Fresca River valley.

Type Subsurface Section. Manzano no. 7, 3,561–4,308 m.

Paleocene

Oazian Stage

Characteristics and Significant Fossils. This stage is also confined to almost the same boundaries as the Manzanian but is somewhat less extensive (Fig. 16). It is well developed in the western part of the basin where it is thickest (more than 700 m) in the Tres Brazos no. 1 well. This stage is composed of hard, dark-brownish shales with a few sandstone beds aggregating 40 m. The top of this stage is marked by the first occurrence of *Spiroplectammina grzybowskii* Frizzell, *Anomalina rubiginosa* Cushman, *Bulimina gonzalezi* Cañon and Ernst, n. sp., *Gyroidina infrafosa* Finlay, *Epistominella texana* (Cushman).

Paleoecology. Marine depth of 1,000–2,000 m.

Type Surface Section. Eastern coast of Brunswick Península near Fundo San Jorge.

Type Subsurface Section. Manzano no. 7, 4,308–4,785 m.

CRETACEOUS-TERTIARY BOUNDARY

Paleocene: Danian

Germanian Stage

Characteristics and Significant Fossils. Although this stage covers much of the basin (Fig. 17), it is poorly developed to absent in the east but is well developed in the west where it is thickest (more than 300 m) in the Kerber no. 1 well. It is composed of hard, dark-greenish shale, silty shale, and shaly sandstone with abundant glauconite. The top of this stage is marked by the first occurrence of *Tritaxia rugulosa* (ten Dam and Sigal), *Bolivina incrassata* Reuss, and *Praeglobobulimina kickapoensis* (Cole).

Paleoecology. Marine depth of 1,000–2,000 m.

Type Surface Section. Vicinity of Blanco and Canelos Sur Rivers on the eastern coast of Brunswick Península.

Type Subsurface Section. Tranquilo no. 2, 2,370–2,619 m.

CRETACEOUS

Upper Cretaceous: Santonian-Maestrichtian

Riescoian Stage

Characteristics and Significant Fossils. This stage extends over most of the basin (Fig. 18) and is thickest (more than 3,000 m) in the Ultima Esperanza area in the west and thins to the vanishing point in the east. It is a turbidite sequence composed of hard, dark, gray-green shale, siltstone, and sandstone exhibiting in the type surface section the sedimentary pattern typical of turbidity current deposition. The top of this stage is marked by the first occurrence of *Globigerina cretacea* d'Orbigny, *Psamminopelta minima* (Cushman and Renz), *Bolivina incrassata* Reuss var. *gigantea* Wicher, *Spiroplectammina gutierrezi* Cañon and Ernst, n. sp., *Bolivinoides draceo dorreeni* Finlay, and *Heterohelix globulosa* (Ehrenberg).

Paleoecology. Probable marine depth of 1,000-2,000 m.

Type Surface Section. Rocallosa Point and Fuentes Bay on the north coast of Riesco Island.

Type Subsurface Section. Vania no. 1, 2,643-2,964 m, and Pampa Larga no. 1A, 1,917-2,026 m.

Upper Cretaceous: Cenomanian-Santonian

Lazian Stage

Characteristics and Significant Fossils. This stage covers the entire basin (Fig. 19) and is thickest (more than 2,000 m) in the Ultima Esperanza area. It, too, is a turbidite sequence of dark-gray shale, siltstone, and sandstone. Thick gravitites of conglomerate with a mudstone matrix were deposited in the Ultima Esperanza area by enormous subsea gravity flows, which cut huge grooves in the underlying soft shale. The resulting groove casts indicate flow direction from northwest to southeast, which was also the flow direction of the turbidity currents. The top of this stage is marked by the first occurrence of *Cibicidoides semiumbilicatus* Toutkovski, *Globotruncana chileana* Cañon and Ernst, n. sp., *Globigerina wenzeli* Cañon and Ernst, n. sp., *Planulina popenoi* (Trujillo), *Globotruncana* (*Globotruncana*) *marginata* (Reuss), and very abundant *Inoceramus* prisms.

Paleoecology. Marine depth of 1,000-2,000 m. May be deeper than the Riescoian.

Type Surface Section. Toro Hill in northern Ultima Esperanza.

Type Subsurface Section. Pampa Larga no. 1A, 2,026-2,253 m.

Lower Cretaceous: Albian-Cenomanian

Peninsulian Stage

Characteristics and Significant Fossils. This stage extends over the entire basin (Fig. 20) and is thickest (400 m) in the western foothills of the Ultima Esperanza

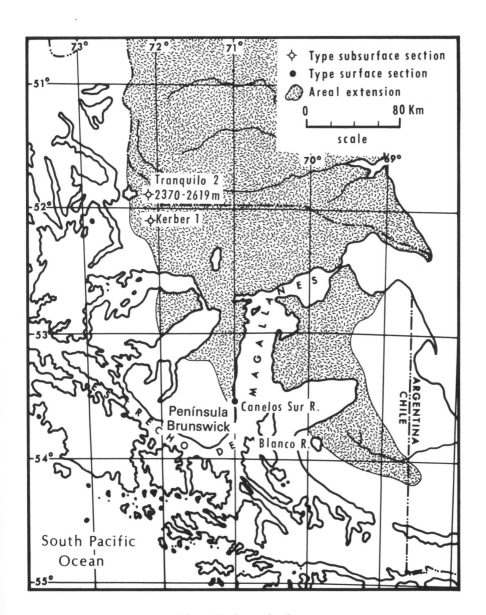

Figure 17. Germanian Stage

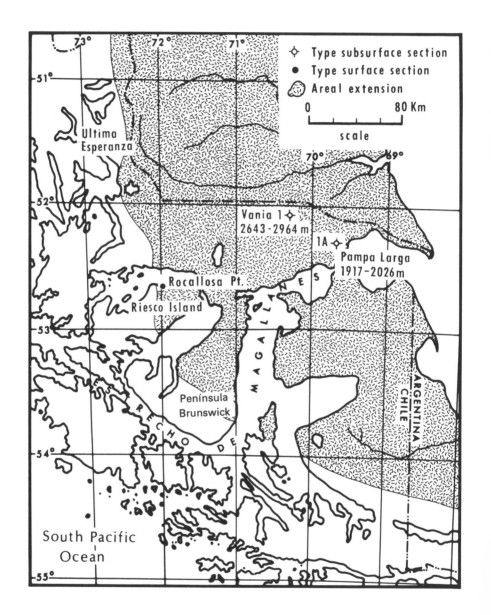

Figure 18. Riescoian Stage

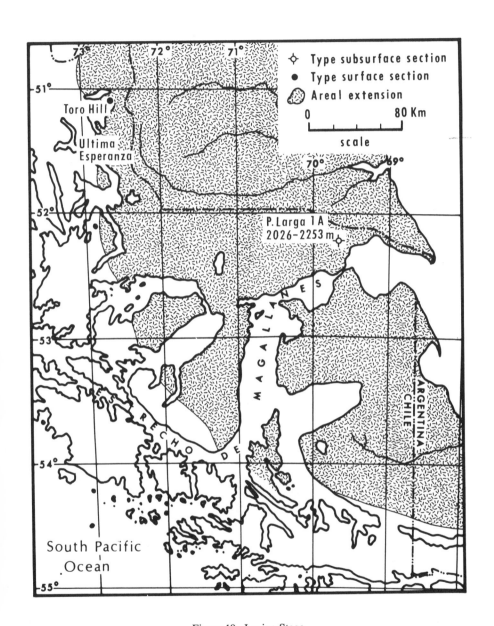

Figure 19. Lazian Stage

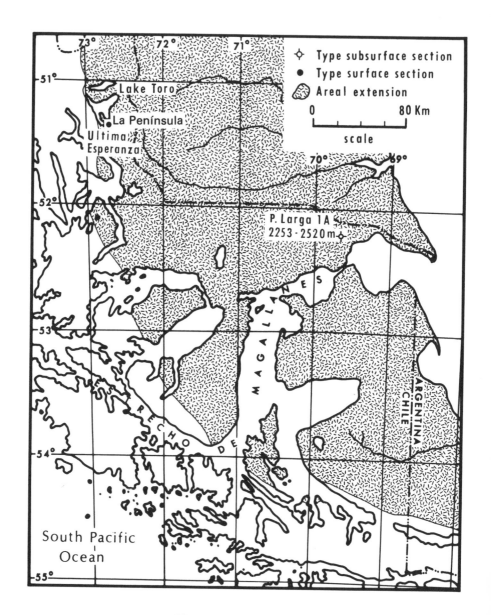

Figure 20. Peninsulian Stage

area. Faulting and folding have greatly disturbed this stage, making thickness calculation for the surface section very difficult. The Peninsulian is composed of dark-gray shale. The top of this stage is marked by the first occurrence of *Tritaxia porteri* (Cañon and Ernst, n. sp., *Hastigerina escheri escheri* (Kaufmann), *Hedbergella planispira* (Tappan), *Spiroplectinata annectens* (Parker and Jones), *Dorothia mordojovichi* Cañon and Ernst, n. sp., *Heterohelix moremani* (Cushman), and Radiolaria *Spumellaria*? sp. 2 (black). For some environmental reason Radiolaria *Spumellaria*? sp. 1 changes from white to black and for charting purposes is called sp. 2.

Paleoecology. Marine depth of 1,000–2,000 m.

Type Surface Section. La Península on the south side of Lake Toro in Ultima Esperanza.

Type Subsurface Section. Pampa Larga no. 1A, 2,253–2,520 m.

Lower Cretaceous: Aptian-Albian

Tenerifian Stage

Characteristics and Significant Fossils. This stage covers the entire basin (Fig. 21) and is thickest (1,000 m) in the Toro no. 2 well in the Ultima Esperanza area. It is composed of light-gray and reddish, limy shale with some white shaly lime. This stage differs lithologically from those above and below by having a large percentage of carbonate that probably resulted from a slower rate of clastic deposition. However, in the lower part of the Tenerifian, clastic sedimentation predominated, and phthanitic, speckled shale was deposited. The top of this stage is marked by the first occurrence of *Pullenia natlandi* Cañon and Ernst, n. sp., *Cibicides* cf. *djaffaensis* Sigal, *Discorbis minima* Vieaux, algal deposits, and Radiolaria *Spumellaria*? sp. 4 (pink and white). The pink color results from staining by pink carbonate in which the radiolarians are found. When the carbonate is white, the radiolarians are white.

Paleoecology. Estimated marine depth of 50–300 m. In general, the foraminiferal fauna suggests 300 m of marine water. The presence of algal deposits indicates that enough sunlight to support algal growth reached the bottom. Usually, in clear water there is not sufficient sunlight below 50 m for plants to grow. However, this algal material may not be indigenous but may have been transported downslope from shallow water into deep water. The widespread uniform thickness of this carbonate member suggests a shelf environment.

Type Surface Section. Tenerife hill in western Ultima Esperanza.

Type Subsurface Section. Pampa Larga no. 1A, 2,520–2,802 m.

Lower Cretaceous: Barremian

Pratian Stage

Characteristics and Significant Fossils. This stage is very widespread (Fig.22) and is thickest (200 m) in the Ultima Esperanza area. Like the Peninsulian, it has much faulting and folding in the foothill belt. It is composed of hard, dark-gray, laminated shale. The top of this stage is marked by the first occurrence of *Lenticulina reyesi* Cañon and Ernst, n. sp.

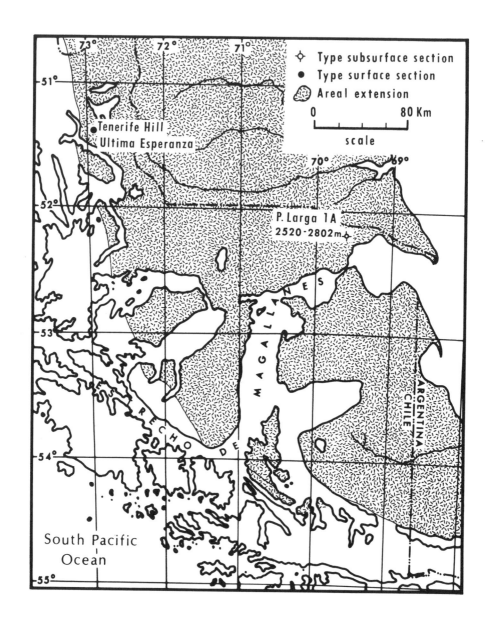

Figure 21. Tenerifian Stage

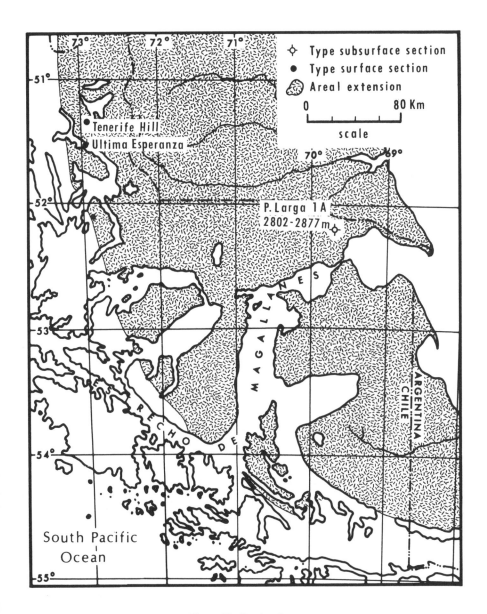

Figure 22. Pratian Stage

Paleoecology. Marine depth of 100-500 m. The microfauna suggests a deeper depositional environment than that of the Tenerifian.
Type Surface Section. Central part of Tenerife hill in western Ultima Esperanza.
Type Subsurface Section. Pampa Larga no. 1A, 2,802-2,877 m.

JURASSIC-CRETACEOUS

Upper Jurassic-Lower Cretaceous: Portlandian-Hauterivian

Esperanzian Stage
Characteristics and Significant Fossils. This stage covers the entire basin (Fig. 23) and is thickest (400 m) in the Ultima Esperanza area. It is composed of dark, brownish-gray shale. The top of this stage is marked by the first occurrence of *Polymorphina martinezi* Cañon and Ernst, n. sp., *Ammobaculites barrowensis* (Tappan), *Astacolus microdictyotos* Espitalie and Sigal, *Lenticulina besairiei* Espitalie and Sigal, *Lenticulina biexcavata* (Myatliuk), and *Haplophragmium inconstans* var. *erectum* Bartenstein and Brand.
Paleoecology. Marine depth of 100-500 m.
Type Surface Section. Rincon River-Tenerife hill area in western Ultima Esperanza.
Type Subsurface Section. Pampa Larga no. 1A, 2,877-3,009 m.

JURASSIC

Upper Jurassic: Oxfordian-Kimmeridgian

Rinconian Stage
Characteristics and Significant Fossils. This stage extends over the entire basin (Fig. 24) and is thickest (500 m) in the Fontaine River area. It is composed mainly of massive quartzitic sandstone, siltstone, and silty shale. The top of this stage is marked by the first occurrence of these microfossils: *Astacolus tricarinellus* (Reuss), *Astacolus filosa* (Terquem), *Marginulinopsis lituoides* (Borneman), *Hoeglundina porcellanea* (Brückmann), *Astacolus stillus* (Terquem), *Dentalina soluta* (Reuss), *Vaginulinopsis ectypa* Loeblich and Tappan, *Reinholdella* cf. *quadrilocula* Subbotina and Datta, *Reinholdella fuenzalidai* Cañon and Ernst, n. sp., Radiolaria *Spumellaria*? sp. 6, and Radiolaria *Nassellaria* sp. 3. These megafauna are also present: *Favrella americana* (Favre), *Favrella steinmanni* (Favre), *Belemnopsis patagoniensis* Favre, *Lucina* cf. *neugoensis* Haupt, *Apticus* sp., *Camptonectes* sp., *Aulacosphinctes* sp., *Trigonia* sp., and *Gryphaea* sp. These flora also occur: *Otozamites sanchtaecrucis* Feruglio, *Sphenopteris patagonica* Halle, and *Gleichenites* cf. *san martini* Halle.
Paleoecology. Depositional environment varying from marine, with a depth of 1-200 m, to nonmarine. Microfauna in the Vania no. 1 suggest water 100-200 m deep.

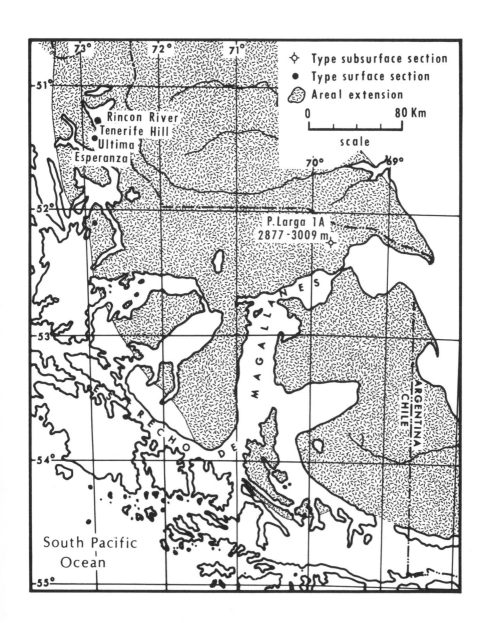

Figure 23. Esperanzian Stage

Type Surface Sections. Bellota Creek in western Ultima Esperanza, Monton Point on the south coast of Seno Almirantazgo, and Fontaine River in southern Tierra del Fuego.

Type Subsurface Sections. Vania no. 1, 3,801-3,940, and Pampa Larga no. 1A, 3,009-3,087 m.

Petroleum in commercial quantities is produced only from Springhill sands that occur in the Rinconian Stage.

Tobifera Series

Characteristics and Significant Fossils. This series has the greatest areal extent of all the stages discussed (Fig. 25). It is composed of tuffaceous volcanic rocks, welded tuff, and ash beds in the Chilean part of the Magallanes Basin. Some *Crustacea*, *Anura*, and flora are reported from this series in Argentina. Its greatest known thickness is 2,176 m in the Cullen no. 64.

Paleoecology. Mostly of continental origin with some shallow-water accumulation in the Argentina part of the basin.

Type Surface Section. Rincon River valley.

Type Subsurface Sections. Maria Emilia no. 3, 1,734-2,785 m, and Cullen no. 64, 1,714-3,890 m.

PALEOZOIC

Paleozoic Series. Metamorphic basement (Figs. 26, 27).

Characteristics. Eroded surface of schist, gneiss, and granodiorite gneiss. The basement rock complex of the Magallanes Basin.

Type Surface Section. Monte Buckland-Fiordo Almirante Martinez area, southwestern tip of Tierra del Fuego.

Type Subsurface Section. María Emilia no. 2, 1,836-1,942 m.

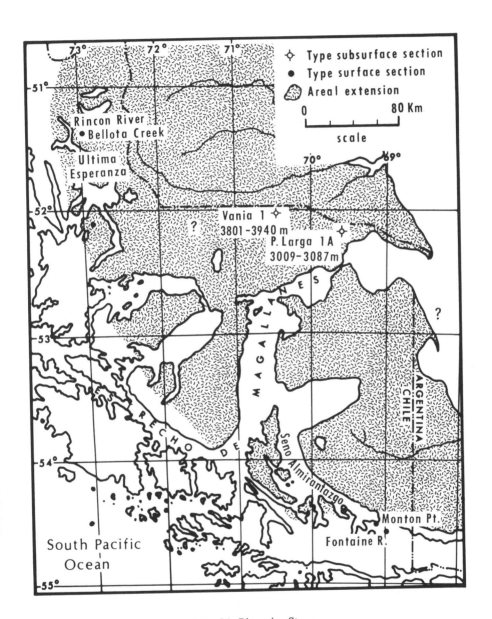

Figure 24. Rinconian Stage

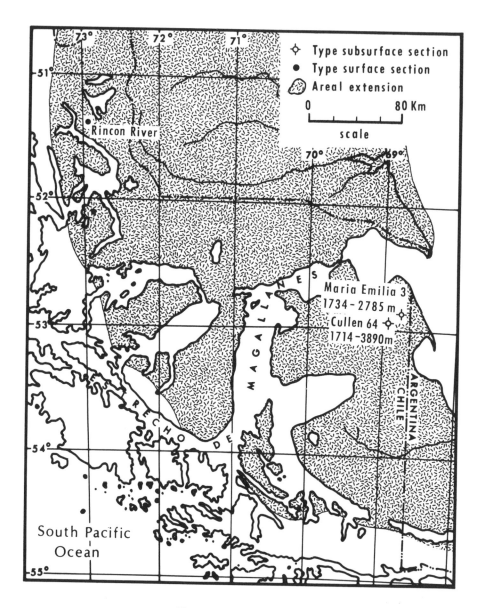

Figure 25. Tobifera Series

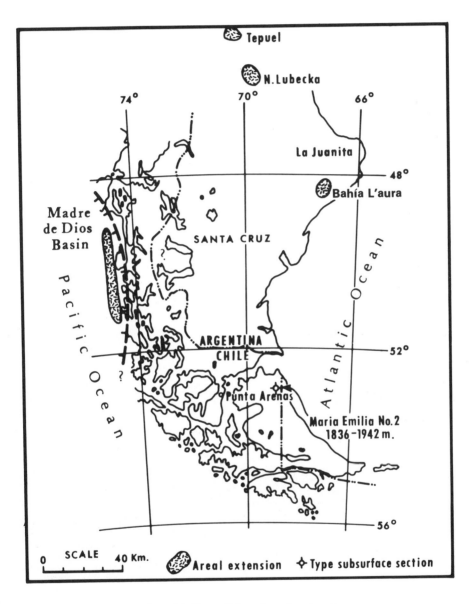

Figure 26. Upper Paleozoic rock outcrops

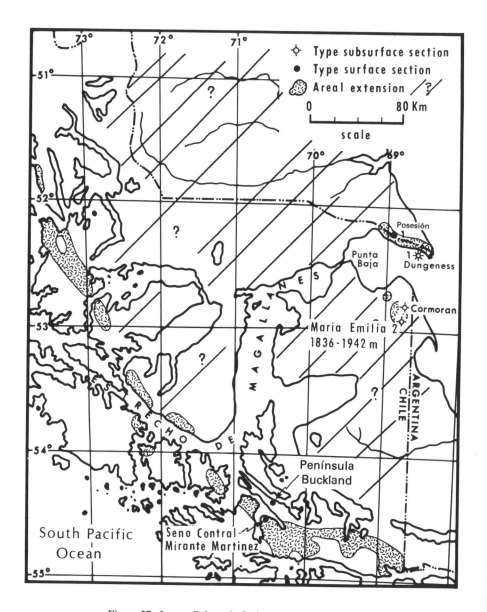

Figure 27. Lower Paleozoic Series (crystalline basement)

Correlation with Other Areas

The Magallanes Basin is so remote from type areas of the European stages that such correlation is difficult. In view of the partial tendency toward biogeographic provincialism observed in this basin, exact equivalence of Magallanes Stages with those in other regions is not claimed. The best hope for correlation may come through linking the Magallanes sequence with central Chile and other areas around the Pacific.

Three lines of approach for correlation have been attempted: (1) correlation of Magallanes Tertiary with central Chile; (2) correlation of Magallanes Tertiary with Peru, Colombia, New Zealand, and California; and (3) correlation of Magallanes Upper Cretaceous with northwest Peru, Texas, Trinidad, New Zealand, and Australia.

The Magallanes species of direct significance, contained in the American continent and overseas correlation, and the local stage subdivisions, with special emphasis on pelagic Foraminifera, are shown on Table 2. The relationships of the Magallanes sedimentary sequence microfauna to other areas in America and overseas, and their age determinations, have been established previously by the following authors: (1) Todd and Kniker (1952), Agua Fresca formation fauna correlation with some upper Eocene fauna of California; (2) Kniker (1947, 1949), Magallanes Tertiary fauna age determination and correlation with California, Peru, and Trinidad, Eocene to Miocene fauna; (3) Robles and others (1956), Magallanes Upper Cretaceous and Paleocene correlation with Mexico and Texas uppermost Cretaceous fauna; (4) Martinez and Ernst (1960), Jurassic-Cretaceous Magallanes sequence age determination; (5) Martinez and others (1964), Magallanes upper Tertiary fauna correlation with California; (6) Sigal (1967), chronostratigraphic of Upper Jurassic-lowermost Cretaceous Magallanes sequence and correlation with Madagascar; (7) Cañon and Ernst,[1] Magallanes Miradorian Stage correlation with central Chile lower Huilma Formation; (8) Barwick (1955) and Charrier (1968), Arauco province (central Chile) marine lower Tertiary correlation with Magallanes Agua Fresca formation fauna. In addition to the above-mentioned works, the following bibliography has been used for correlation: central Chile: Martinez (1968), Sociedad Geologica de Chile (1968); Peru: Weiss (1955); Trinidad: Brönnimann (1952); New Zealand: Hornibrook (1958); Australia: Edgell (1957); Texas: Frizzell (1954), Cushman (1951); California: Mallory (1959); Colombia: Petters and Sarmiento (1956); and Madagascar: Espitalie and Sigal (1963).

---

[1] ENAP—Foraminiferal chart, Huilma no. 1 well.

Paleoecologic-Sedimentary Summary and Conclusions

Plotting graphically the total fauna found in each well and outcrop sample, and coupling this information with detailed stratigraphic descriptions together with electric log data, has provided the information necessary to construct a stage system based on time.

This method of investigation provides a better understanding of the lateral variation of faunas in a sedimentary basin and greatly reduces the difficulties usually encountered in regional correlation of sedimentary sequences.

A summary of the stage system and correlation in Magallanes Basin is shown in Table 1.

The oldest stage recognized is the Rinconian, which was deposited in marine depths ranging from 1 to 200 m, with the water deepest in the central part of the basin (Vania no. 1 well) and gradually shallowing toward the east in the Cullen area. During Esperanzian time the water deepened to range from 100 to 500 m. Water depth appears to have shallowed during the Tenerifian Stage, as suggested by the abundance of molluscan faunas and algal remains.

From Peninsulian through Cameronian time, the sea ranged in depth from 1,000 to 2,000 m, as indicated by such deep water forms as *Pullenia bulloides, Praeglobulimina subcalva, Cyclammina cancellata, Bulimina corrugata (rostrata)* and abundant *Radiolaria*. Apparently, the position of isostatic balance was maintained at nearly 1,500 m through this range. However, from Rosarian time to the present the rate of subsidence was exceeded by the rate of sedimentation, with the result that the basin was filled to sea level.

Two cross sections have been prepared showing the correlation based on time-stratigraphic units (Tables 3 and 4). The section on Table 3 extends from the Evans no. 1 well to the Vania no. 1. Here, during the Gaviotian stage, the water shallows from a depth of about 1,000 m in the Cisne no. 1 to a littoral environment abundant with only mollusk fragments in the Vania no. 1.

The same phenomena are shown on Table 4, which extends from the María Emilia no. 2 well to the Kerber no. 1. Here the Miradorian Stage shows a lateral facies variation which indicates a deepening of the basin from the Kerber no. 1 toward the Rio del Oro-Sombrero area.

Fortunately, excellent sample suites with good faunas, together with electric logs from closely spaced wells, make it possible to establish definitely the changes in water depth within the same time zone.

The use of recently improved ocean bottom sampling equipment has considerably extended and clarified our knowledge of sedimentary processes and conditions in present deep-sea basins. An examination by Natland of samples collected by Scripps Institution of Oceanography, La Jolla, California, from the San Diego trough reveals that the greatest thickness of clean, well-sorted sand is found in the basal parts of the trough and that, in general, only a thin layer of silt and clay is being deposited on the sides of the trough.

This information is very important to the study of the Magallanes Tertiary Basin because of its similarity to the San Diego trough in shape and in its relationship to the shoreline. In view of such marked similarity, it is reasonable to expect to find the greatest thickness of good permeable sand suitable for oil reservoirs in the central part of the Magallanes Basin. It is also probable that the Magallanes Basin was sufficiently deep to provide the gradients necessary to permit traction and gravity flows of clean sand to the deeper part of the trough.

Another interesting sedimentary condition is shown in Table 3. In the area of the Evans and Miraflores wells, during Moritzian through Brunswickian time, a great thickness (1,732 m in the Evans no. 1) of sandstone and conglomerates (Ballena Formation) was probably deposited by sedimentary gravity flows and turbidity currents. These coarse clastics lense out updip between the Evans no. 1 and the Rio Chico no. 1 wells. It should be noted that, to transport coarser clastics in suspension for great distances, a turbidity current mixture must include sufficient (usually about 30 percent) fine material to provide a matrix capable of prolonging the suspended state of the coarser particles so that they continue to flow in response to gravity. Unless the matrix in the graded sand part of a turbidite has been sufficiently winnowed by ocean bottom currents to clean the sand appreciably, the turbidite sandstone will probably have insufficient porosity to allow oil to migrate and will, therefore, not be a good reservoir. Consequently, the attractiveness of this area for oil exploration depends on the amount of matrix in the sandstone and its effect on porosity.

From an oil exploration standpoint, one of the most important features revealed by this paleoecologic-sedimentary study of the Magallanes Basin was the presence of several sandy time-stratigraphic units in the central part of the Tertiary basin, whereas they are either poorly developed or absent on the eastern and western sides. These units are the Oazian, Manzanian, Brunswickian, Moritzian, Cameronian, and Rosarian Stages shown in Tables 3 and 4. Paleoecology strongly indicates that these deep-water sandy sediments in the center of the Magallanes trough came from a western source, the Main Cordillera area and western Santa Cruz province, where thick Upper Cretaceous sandstones were being eroded and swept into a submarine canyon down which they were moved by gravity, traction, and (or) turbidity current flows to the bottom of the trough. Considerable winnowing of matrix-filled turbidite sands by strong basin bottom currents can occur during traction transport, which produces a clean permeable sand very suitable for oil reservoirs.

Acknowledgments

This study was made possible through the support and permission given by the Empresa Nacional del Petróleo (ENAP-Chile). The authors are grateful for the enthusiastic encouragement, in all phases of this work, of Osvaldo Wenzel, exploration manager of ENAP in Santiago, and of Oscar Schneider, manager of the ENAP Exploration Department in Punta Arenas.

The cooperation of Antonio Cañon, senior micropaleontologist, and Mario Ernst, assistant micropaleontologist, and all of the personnel of the Micropaleontological Laboratory, is gratefully acknowledged.

Thanks are extended to Raul Cortes and Agustin Gutierrez, who offered suggestions on the manuscript.

The writers are also indebted to numerous geologists and paleontologists, who have carried out geological studies in southern Patagonia, for the data used to elaborate this report.

Part II
Magallanes Basin Foraminifera

Abstract

The microfaunas in 42,000 samples from outcrop sections and more than 60 key wells in the Magallanes Basin were plotted graphically at about 3-m stratigraphic intervals. Three hundred and fifty species of Foraminifera belonging to 118 genera have been recognized. Thirteen species considered to be new are illustrated on six plates. In addition, five species of Radiolaria have been found to be useful. Mollusks, Inoceramus prisms, fish remains, ostracods, diatoms, carbonaceous matter, glauconite, and volcanic ejecta are also useful for correlation.

The method used in sample preparation and in plotting graphically the fossils found in each sample is also described. A foraminiferal distribution chart of the ENAP Sombrero no. 1 well and the Vania no. 1 well is included (Table 5). Work by previous authors is briefly reviewed.

Previous Work

The earliest known work on Foraminifera in Patagonia and southern Chile was that of Brady (1884). This work is of primary importance to matters related to the geographic and bathymetric distribution of existing species and varieties of Foraminifera. Many of the living species have identical and very closely related forms extending back into the Cretaceous, which provides an excellent guide to paleoenvironments.

Two subsequent papers by Richter (1925) and Kranck (1932, 1933) on Tierra del Fuego and the southern islands are of interest in that they describe Radiolaria from the Lower Cretaceous sedimentary rocks.

The pioneer students of southern Chile Foraminifera were J. S. Hollister, C. Mordojovich, and H. T. Kniker, who dealt mainly with Tertiary and Upper Cretaceous species. Hollister made the first micropaleontological investigations. Mordojovich began the classification of Foraminifera from Brunswick Peninsula and showed the enormous practical value of micropaleontology in oil exploration in the Magallanes Basin. But it was Hedwig T. Kniker who organized an effective micropaleontological laboratory staff for ENAP, which carried on the work in Punta Arenas from 1945 to 1950.

Many reports on Foraminifera have been written since that date, most of them unpublished. Todd and Kniker (1952) described an important Eocene assemblage from the Agua Fresca formation; the type locality is located south of Punta Arenas. This paper contains the only monographic description available of one of the most prolific Eocene fauna found in south Patagonia. Later, between the years 1950 and 1968, a series of unpublished reports were written for ENAP by E. Severin, W. O'Gara, R. Martinez, M. Ernst, A. Gutierrez, A. Cañon, and M. L. Natland. These reports contain descriptions of the microfaunas used for correlation and age determination of outcrop and well sections throughout the Magallanes Basin.

Robles and others (1956) discussed the stratigraphic and paleoecological conditions of Fuentes, Rocallosa, Rio Blanco, and Chorrillo Chico formations of the Cretaceous-Tertiary boundary in the foothill zone. They also found many previously published names for Foraminifera occurring in the lithostratigraphic units mentioned above.

Some micropaleontological data from the numerous unpublished reports of ENAP, Chile, were utilized by Hoffstetter and others (1957).

Martinez and others (1964), in a detailed unpublished report, described the Foraminifera from the Ciervos Formation type locality south of Punta Arenas. One noteworthy subsequent paper, by Martinez (1964), reported the occurrence of *Bolivinoides draceo dorreeni* Finlay in the lower part of the Rocallosa formation at the type locality situated in the foothill belt. He dated this unit and discussed the paleoecologic conditions toward the end of Maestrichtian time.

Between 1961 and 1966, M. L. Natland applied the foraminiferal charting system to provide more detailed information necessary for establishing a stage system for correlating Magallanes Basin Tertiary and Cretaceous sediments. Most of the Foraminifera examined came from cores and ditch samples. His views are summed up in a long series of unpublished reports for ENAP, Chile.

The Cretaceous-Tertiary boundary problem in Magallanes Basin is discussed in two papers by Hauser (1964) and Charrier and Larsen (1965).

A review of the present state of knowledge, dealing with the Cretaceous and Tertiary sedimentary sequences and the microfauna of the Magallanes Basin, was presented by Herm (1966).

Cañon (1968) adapted the Natland-Gonzalez system of stages for correlation of the Tertiary sedimentary sequence of Tierra del Fuego. He dealt largely with microfaunas collected from outcrops complimented with ditch samples from many explorations.

Recently, Sigal (1967) in a preliminary report, carried out a chronostratigraphic study of the "Estratos con Favrella" and Springhill formations.

There are very few publications on the micropaleontology of southern Argentina. The first reports on the Santa Cruz provinces and Tierra del Fuego were by Camacho (1957). He described a Paleocene fauna from the Argentine part of Tierra del Fuego. In connection with petroleum exploration, micropaleontology research was carried out by Shell Oil Company and others. Garcia and Camacho (1965) described the microfauna from an exploratory well in the Santa Cruz province. Malumian (1969) described the Foraminifera from the Upper Cretaceous-Tertiary sequence of the Santa Cruz no. 3 well.

Magallanes Basin Microfauna

The microfaunas described in this paper were extracted from well cuttings, cores, and outcrop sections by the authors and other personnel from ENAP (Empresa Nacional del Petróleo-Chile).

About 42,000 samples were processed and their microfaunas charted from outcrop sections and from more than 60 wells. Most cuttings were taken at 3-m intervals by carefully trained well-site geologists.

The uncontaminated condition of these samples is evident from the faunal distribution patterns on the charts. Foraminiferal charting registers a sharp lower limit to an abundant foraminiferal zone, which has persisted in the cuttings through considerable well footage, thus indicating the lack of cavings in ditch samples examined.

To process the large number of samples studied for microfaunal examination, the following procedures were followed:

1. Identification and labeling of samples as to location was done at the well and in the laboratory.
2. A mechanical crusher was used to break the sample material into fragments less than 0.5 in. in diameter.
3. A 50-cm^3 sample was taken for treatment.
4. Disaggregation of rather indurated shales and siltstone was accomplished by: (1) soaking the sample in kerosene 0.5 to 1 hr; (2) washing, wet sieving, and soaking the sample with hot water; (3) placing samples and water in jars containing a 1-1/4 in. diameter rubber-coated iron pestle and rolling jars on motor driven rollers 0.5 to 12 hr (this reduces the shale and siltstone to fine mud, freeing microfossils); and (4) wet sieving the sample through a 200-mesh screen. Softer shales and siltstone samples were separated by (1) soaking and boiling the sample with a solution of sodium bicarbonate 1 to 2 hr, and (2) washing and screening the sample with a 200-mesh screen. Most specimens, especially in the older formations, were internally filled with various materials and therefore could not be separated from rock fragments with heavy liquids.
5. Residues were dried on a hot plate or in an oven and screened through nested screens of from 14 to 200 mesh.
6. Residues were placed in labeled envelopes, ready for examination. Microfaunal species and some minerals in each sample were plotted graphically with

the aid of a scroll-type plotting system invented by Natland. This system contains two rollers on which blank cross-section paper (1 in. × 1 in. × 20 in. wide) is spooled. The paper passes from the roller, on which the cross-section paper is spooled, over a 6-in. writing surface where the relative abundance of species can be recorded. This is accomplished by using a symbol system (Natland, 1937, unpub. data) which can be progressively elevated in value, without erasure, as the sample is studied. The species abundance is estimated with the aid of a 4-in. diameter glass plate with a circle on one side divided into 50 equal segments. The washed sample is spread radially and evenly about 0.5 in. from the maximum diameter of the plate. This permits abundance estimation by examining only fractional parts of the sample. The faunal abundance data is plotted on the same scale as the electric log with added lithologic data (Table 1). This greatly facilitates accurate correlation and readily displays faunal variations resulting from ecologic changes within the same time-stratigraphic unit.

In this investigation, about 150 genera and more than 350 foraminiferal species of Tertiary, Cretaceous, and Jurassic age have been recognized.

Two faunal lists, here referred to as headers, contain the foraminiferal species used for correlation. Under these headers, the relative abundance of each species in each sample is plotted.

The first header contains 200 Tertiary species. The second has 150 species for recording the distribution of Cretaceous and Jurassic species.

Table 1 contains a composite range chart of the Sombrero no. 1 well and the lower sedimentary section (Estratos con *Favrella steinmanni* and Springhill formations) in Vania no. 1. The range of species that has proven to be the most useful for correlation is shown on this chart. However, the maximum basin-wide stratigraphic ranges of many species are more extensive than indicated on the chart of these two wells.

There are many advantages to this charting system. It provides a rapid, currently developing, graphic view of faunal distribution. Also, from sample to sample, one can immediately observe zone development and omissions in recording.

Without the foraminiferal distribution visible on these charts it would be difficult, if not impossible, to formulate the stage system presented in Part I. Prior foraminiferal correlations made in this area, without the use of this plotting system, were sometimes off by several hundred feet.

In general, planktonic Foraminifera are too poorly represented in the Tertiary of the Magallanes Basin to be useful for correlation and age determination. The upper Mesozoic contains abundant planktonic Foraminifera that are excellent for correlation.

Systematic Micropaleontology

More than 350 foraminiferal species have been recognized throughout the Cretaceous and Tertiary of the Magallanes Basin. Of these, only 74 species diagnostic of their respective stage levels are described below. The systematic arrangement of these species follows the Loeblich-Tappan format set forth in the *Treatise on Invertebrate Paleontology* (1964).

Holotypes of the new species are deposited in the Museo Nacional de Historia Natural, Santiago, Chile. Paratypes are filed in the U.S. National Museum, Washington, D.C.

Order FORAMINIFERIDA
Suborder TEXTULARIINA Delage and Herouard, 1896
Superfamily LITUOLACEA de Blainville, 1825
Family RZEHAKINIDAE Cushman, 1933

Genus *Psamminopelta* Tappan, 1957
Psamminopelta minima (Cushman and Renz)
(Pl. 1, figs. 1a, b)

Rzehakina epigona (Rzehak) var. *minima* Cushman and Renz, 1946, p. 24, Pl. 3, fig. 5.

Remarks. Our specimens closely resemble *R. epigona* var. *minima*, which we have reassigned generically to *Psamminopelta* because the specimen illustrated by Cushman and Renz is evolute rather than involute. Since we do not regard *minima* as being related to *R. epigona*, we have raised it to specific rank. Our specimens are not sufficiently sigmoid in cross section to place in *Spirosigmoilinella*. The chamber arrangement differs from *minima* in that the chambers are not added symmetrically along a median line but are skewed considerably from center at the base.

Distribution. Riescoian Stage.
Locality. Sample G-421, Bahía Fuentes, Brunswick Península.

Psamminopelta venezuelana (Hedberg)
(Pl. 1, figs. 2a, b, 3)

Rzehakina venezuelana Hedberg, 1937, p. 669, Pl. 90, fig. 12.

Remarks. Hedberg figures a megalospheric form identical to our megalospheric specimens, but we have reassigned this evolute species to *Psamminopelta* instead of leaving it in involute *Rzehakina*. Our microspheric form (Pl. 1, fig. 2a) tends to add initial chambers in a sigmoid fashion while our megalospheric form (Pl. 1, fig. 3) adds chambers radially.
Distribution. Rare to abundant in Miradorian through Cameronian Stages.
Locality. ENAP Josefina no. 1, 421.5-425 m.

Family LITUOLIDAE de Blainville, 1825
Subfamily CYCLAMMININAE Marie, 1941

Genus *Cyclammina* Brady, 1879
Cyclammina cancellata Brady
(Pl. 1, figs. 4a, b)

Cyclammina cancellata Brady 1897. Cushman, 1950, Pl. 6, figs. 3, 4.

Remarks. Typical *Cyclammina cancellata* are widely distributed throughout the middle and lower Tertiary of the Magallanes Basin and range through most of the Cretaceous in the Pampa Larga area.
Distribution. Gaviotian through Peninsulian Stages.
Locality. ENAP Leña Dura no. 1, 891-900 m.

Subfamily LITUOLINAE de Blainville, 1825

Genus *Ammobaculites* Cushman, 1910
Ammobaculites barrowensis Tappan
(Pl. 1, figs. 5a, b)

Ammobaculites barrowensis Tappan, 1955, p. 45, Pl. 11, figs. 7-12.

Remarks. Specimens from the Magallanes Basin appear identical to those described by Tappan from the Arctic type locality. This is very interesting because of the geographic location.
Distribution. Esperanzian Stage.
Locality. ENAP Victoria Sur no. 1, core no. 1, 2,267-2,273 m.

Family TEXTULARIIDAE Ehrenberg, 1838
Subfamily SPIROPLECTAMMININAE Cushman

Genus *Spiroplectammina* Cushman, 1927
Spiroplectammina adamsi Lalicker
(Pl. 1, figs. 6a, b)

Spiroplectammina adamsi Lalicker, 1935. Todd and Kniker, 1952, p. 6, Pl. 1, figs. 18, 19.

Remarks. This distinctive species is widely distributed in the Magallanes Basin.
Distribution. First occurrence marks top of Brunswickian Stage, continues abundantly through the Manzanian Stage.
Locality. ENAP Manzano no. 1, core no. 236, 3,012–3,016.5 m.

Spiroplectammina grzybowskii Frizzell
(Pl. 1, figs. 8a, b)

Spiroplectammina grzybowskii Frizzell, 1943, p. 339, Pl. 55, figs. 12, 13.

Remarks. Ruth Todd sent us Frizzell's paratypes from the U.S. National Museum for examination. These specimens, from the Upper Cretaceous Mal Paso formation of Peru, closely resemble our Upper Cretaceous specimens from the Magallanes Basin. Our specimens have rougher walls that tend to obscure the sutures, but when wet they show a sutural pattern similar to that of Frizzell's specimens. This species is the most diagnostic of the Riescoian Stage.
Distribution. Oazian, Germanian, and Riescoian Stages.
Locality. Sample JC-292, south of Prat Point, south coast of Seno Otway, Brunswick Península.

Spiroplectammina brunswickensis Todd and Kniker
(Pl. 1, figs. 7a, b)

Spiroplectammina brunswickensis Todd and Kniker, 1952, p. 6, Pl. 1, fig. 16.

Remarks. The Todd and Kniker publication contains a good discussion of this distinctive species and its relation to similar forms.
Distribution. Manzanian Stage.
Locality. Sample M-46, El Ganzo River, Brunswick Península.

Spiroplectammina gutierrezi Cañon and Ernst, n. sp.
(Pl. 1, figs. 9a, b)

Description. Test free, elongate, tapering gradually upward from a small rounded base in the microspheric form, megalospheric form less tapered; periphery broadly rounded with irregular indentations; early chambers planispiral, later chambers biserial, chambers and sutures indistinct curving downward from median line; walls agglutinated with siliceous cement; aperture a short and narrow textularian opening at the interior margin of the last chamber.
Measurements. Microspheric form: length 1.10 mm, width 0.30 mm, thickness 0.15–0.20 mm. Megalospheric form: Length 0.75 mm, width 0.25 mm, thickness 0.10–0.12 mm.
Distribution. Riescoian into Tenerifian Stage.
Locality. Sample MX-2024, Rocallosa Point, Riesco Island.
Depository. Holotype: Museo Nacional de Historia Natural, Santiago, Chile, cat. no. SGO. p.m. Pi. 158. Paratype: U.S. National Museum, Washington, U.S.N.M. no. 688428.

Family TROCHAMMINIDAE Schwager, 1877
Subfamily TROCHAMMININAE Schwager, 1877

Genus *Trochammina* Parker and Jones, 1859
Trochammina cf. *inflata* (Montagu)
(Pl. 1, figs. 10a-c)

Nautilus inflatus Montagu, 1808.
Trochammina inflata (Montagu) Loeblich and Tappan, 1964, p. C259, Fig. 173, 1a-c.

Remarks. Our specimens, although flattened somewhat by overburden weight, are nearly identical to Holocene forms which, in the present marine environment, are restricted to lagoons. It is surprising to find such well-preserved Pliocene forms of this fragile species, which is usually destroyed by washing techniques required to process well samples.
Distribution. Miradorian Stage.
Locality. ENAP San Antonio no. 1, core no. 128, 1,344.5-1,348.5 m.

Family ATAXOPHRAGMIIDAE Schwager, 1877
Subfamily VERNEUILININAE Cushman, 1911

Genus *Spiroplectinata* Cushman, 1927
Spiroplectinata annectens (Parker and Jones)
(Pl. 1, figs. 11a-c)

Textularia annectens Parker and Jones, 1863.
Spiroplectinata annectens (Parker and Jones) Loeblich and Tappan, 1964, p. C272, Fig. 182, 1a, b.

Remarks. Our specimens appear identical to those described by Parker and Jones from the Gault of Folkstone and Biggleswade, England. Specimens from the Gault of Germany were figured by Schmitt.
Distribution. Peninsulian into Esperanzian Stage.
Locality. ENAP Pampa Larga no. 1A, 2,352-2,364 m.

Genus *Tritaxia* Reuss, 1860
Tritaxia chileana (Todd and Kniker)
(Pl. 1, figs. 12a, b)

Clavulinoides chileana Todd and Kniker, 1952, p. 11, Pl. 2, figs. 2-4.

Remarks. Following Loeblich and Tappan, we have changed the generic assignment of this species from *Clavulinoides* to *Tritaxia*. This species resembles *Clavulinoides szaboi* (Hantken) but differs from it by having less concave sides.
Distribution. Clarencian and Brunswickian Stages.
Locality. Sample JSB-82, Leña Dura River, Brunswick Península.

Tritaxia porteri Cañon and Ernst, n. sp.
(Pl. 1, figs. 13 a, b)

Description. Test free, elongate; early chambers triserial followed by two or three biserial chambers succeeded by a long uniserial stage; triserial part triangular in cross section, sides concave with subacute angles; uniserial chambers become more rounded as added, sutures indistinct in first few chambers becoming more depressed in the last chamber; walls smooth to rough, agglutinated with siliceous cement; aperture terminal, round to lunate.

Measurements. Length 1.20 mm, width 0.30 mm.

Remarks. This species does not fit clearly into any existing genus because the biserial stage is too short for *Gaudryinella*, too long for *Clavulina*, and not flat enough for *Spiroplectinata*. We have placed it in *Tritaxia* because the triserial part has concave sides with subacute angles, the biserial part is either absent or consists of only one or two pairs of chambers, and the uniserial stage is long with rounded chambers separated by a depressed suture. Since it occurs with *Spiroplectinata annectens*, *T. porteri* may possibly be related in some way to that species. It differs from *Gaudryinella alexanderi* Cushman by having a uniserial stage.

Distribution. Lower Lazian into the Peninsulian Stage. Most abundant in the Peninsulian.

Locality. ENAP Pampa Larga no. 1A, 2,400–2,415 m.

Depository. Holotype: Museo Nacional de Historia Natural, Santiago, Chile, cat. no. SGO. p.m. Pi. 152. Paratype: U.S. National Museum, Washington, D.C., cat. no. 688429.

Tritaxia rugulosa (ten Dam and Sigal)
(Pl. 1, figs. 14a, b)

Clavulinoides rugulosa ten Dam and Sigal, 1950, p. 32, Pl. 2, figs. 8–10.

Remarks. In our specimens, the triangular cross section extends from initial end through a few chambers in the biserial or uniserial part. Some specimens abruptly change chamber arrangement from triserial to uniserial with no keels. Figures by ten Dam and Sigal show the same variation.

Distribution. Germanian Stage.

Locality. Sample V-9, Chorrillo Garcia, southeast coast of Brunswick Península.

Subfamily GLOBOTEXTULARIINAE Cushman, 1927

Genus *Dorothia* Plummer, 1931
Dorothia mordojovichi Cañon and Ernst, n. sp.
(Pl. 1, figs. 15 a, b)

Description. Test trochospiral initially with about three revolutions, later stage biserial with three or more sets of opposing chambers rapidly increasing in size and strongly overlapping; median line twisted; chambers well rounded, inflated, oval in cross section; sutures distinct, depressed; walls finely arenaceous, smooth with waxy luster, agglutinated, probably with siliceous cement (HC1 does not disintegrate test); aperture an elongate interiomarginal slit.

Measurements. Length 0.60 mm, width 0.42 mm, thickness 0.38 mm.

Remarks. There are several previously described species which closely resemble the above form. Of these, *D. inflata* Colom appears to differ only by having a more round cross section and a straighter median line. *D. plummeri* Brotzen is generally more elongate and does not increase as rapidly in width with added chambers. *D. brevis* Cushman and Stainforth is more round in cross section and chambers are more strongly overlapping. *Karreriella marina* Proto Decima and Ferasin also has the same general shape.

Distribution. Lazian into Esperanzian Stage. Most abundant in the Peninsulian Stage.

Locality. ENAP Pampa Larga no. 1A, 2,301–2,310 m.

Depository. Holotype: Museo Nacional de Historia Natural, Santiago, Chile, cat. no. SGO. p.m. Pi. 153. Paratype: U.S. National Museum, Washington, D.C., cat. no. 688430.

Dorothia principensis Cushman and Bermúdez
(Pl. 1, figs. 16 a, b)

Dorothia principensis Cushman and Bermúdez, 1936, p. 57, Pl. 10, figs. 3, 4.

Remarks. This species described from the Eocene of Cuba has also been found in the Eocene Kreyenhagen shale of California, the Yazoo clay of Mississippi, and the Abuillot formation of Haiti.

Distribution. Brunswickian and Manzanian Stages.

Locality. ENAP well P-7, 541–546 m.

Subfamily VALVULININAE Berthelin, 1880

Genus *Plectina* Marsson, 1878
Plectina elongata Cushman and Bermúdez
(Pl. 1, figs. 18a, b)

Plectina elongata Cushman and Bermudez, 1936, p. 58, Pl. 10, figs. 22–24.

Remarks. This distinctive species forms a well-defined zone in the Magallanes Basin. It is similar to *P. eocenica* Cushman from the Eocene at Biarritz, France, and to *P. garzaensis* Cushman and Siegfus from the upper Eocene Kreyenhagen shale of California.

Distribution. Upper part of Moritzian Stage.

Locality. ENAP Manzano no. 1, 2,332–2,341 m.

Genus *Karreriella* Cushman, 1933
Karreriella cushmani Finlay
(Pl. 1, figs. 17a, b)

Karreriella cushmani Finlay, 1940, p. 452, Pl. 63, figs. 38–42.

Remarks. This species closely resembles *K. cylindrica* Finlay but differs from it by having more globose and less embracing chambers.
Distribution. Miradorian through Brunswickian Stage.
Locality. Sample F-305, Bahía Inútil, Tierra del Fuego.

Suborder ROTALIINA Delage and Herouard, 1896
Superfamily NODOSARIACEA Ehrenberg, 1838
Family NODOSARIIDAE Ehrenberg, 1838
Subfamily NODOSARIINAE Ehrenberg, 1838

Genus *Astocolus* de Montfort, 1808
Astacolus microdictyotos Espitalie and Sigal, 1963
(Pl. 1, figs. 19a-c)

Astacolus microdictyotos Espitalie and Sigal, 1963, p. 33, fasc. 32.

Distribution. Pratian and Rinconian Stages.
Remarks. Our specimens are nearly identical to those described by Espitalie and Sigal (1963) which they collected from Upper Jurassic and Neocomian beds in the Majunga basin of Madagascar. This correlation is further strengthened by several other unique species that occur in both basins (Sigal and others, 1970).

Astacolus skyringensis Todd and Kniker
(Pl. 2, figs. 1a, b)

Astacolus skyringensis Todd and Kniker, 1952, p. 14, Pl. 2, fig. 29.

Remarks. This unique species is characterized by about 12 parallel costae which are continuous across the sutures.
Distribution. Manzanian Stage, with a few occurrences in Lazian and Peninsulian Stages.
Locality. Sample Ruby-4, south coast Seno Otway, Brunswick Península.

Astacolus stillus (Terquem)
(Pl. 2, figs. 2a, b)

Cristellaria stilla Terquem, 1866, p. 517, Pl. 22, fig. 7.

Remarks. Todd and Kniker, 1952, figured a similar form which they called *Astacolus* sp. cf. *crepidula* (Fichtel and Moll). This species was originally described by Fichtel and Moll in 1798 as *Nautilus crepidula* from Holocene sediments off the coast of Italy. *Cristellaria stilla* Terquem was described from Jurassic sediments or at approximately the same level as our Magallanes Basin specimens. There are many other references which show figures very similar to *A. stillus*
Distribution. Rinconian Stage.
Locality. ENAP Sombrero no. 1, 2,192-2,198 m.

Astacolus tricarinellus (Reuss)
(Pl. 2, figs. 3a, b)

Cristellaria (Cristellaria) tricarinella Reuss, 1863, p. 68, Pl. 7, fig. 9.

Remarks. This form has nearly parallel sides with sutures raised to become sharp costae; periphery with three keels; walls have slight reticulation.
Distribution. Lazian into Rinconian. Most abundant in the Peninsulian Stage.
Locality. ENAP Cullen no. 6, core no. 1, 1,701–1,705 m.

Genus *Lenticulina* Lamarck, 1804
Lenticulina cf. *asperuliformis* (Nuttall)
(Pl. 2, figs. 4a, b)

Cristellaria asperuliformis Nuttall, 1930, p. 282, Pl. 23, figs. 9, 10.

Remarks. The degree of ornamentation on this species is variable, and sometimes its thick limbate sutures interlace with longitudinal costae and tubercles. Ornamentation usually decreases as chambers are added. It typically has a sharply keeled periphery. This species has been reported from the Eocene Kreyenhagen shale of California by Cushman and Siegfus and from the lower Eocene in the Tampico region of Mexico by Nuttall.
Distribution. Clarencian Stage.
Locality. Sample JSB-104, Tres Brazos River, Brunswick Península.

Lenticulina biexcavata (Myatliuk)
(Pl. 2, figs. 5a, b)

Cristellaria biexcavata Myatliuk, 1939, p. 56 (Russian) or p. 72 (English), Pl. 4, figs. 41, 42.

Remarks. The Magallanes Basin specimens are nearly identical to that figured by Myatliuk (1939), and they occur at about the same age level.
Distribution. Esperanzian Stage.
Locality. ENAP Chañarcillo no. 4, core no. 2, 2,259–2,262 m.

Lenticulina reyesi Cañon and Ernst, n. sp.
(Pl. 2, figs. 6a, b)

Description. Test free, planispiral, elongate, moderately compressed, vitreous; early chambers closely coiled, last two or three chambers uncoiled; chambers distinct, separated by raised, limbate costae extending beyond the periphery to form nodes, several small nodes in umbilical area; periphery subacute, aperture radiate at top of last chamber.
Measurements. Length 0.07 mm, width 0.45 mm, thickness 0.38 mm.
Distribution. Pratian through Rinconian Stages.
Locality. ENAP Sombrero no. 3, core no. 1, 2,195–2,199 m.
Depository. Holotype: Museo Nacional de Historia Natural, Santiago, Chile, cat. no. SGO. p.m. Pi. 154. Paratype: U.S. National Museum, Washington, D.C., cat. no. 688431.

Genus *Marginulina* d'Orbigny, 1826
Marginulina knikerae Cañon and Ernst, n. sp.
(Pl. 2, figs. 7a, b)

Description. Test large, nearly cylindrical throughout; early chambers incompletely coiled, later chambers rectilinear; sutures slightly oblique, distinct, flush with surface in early chambers becoming depressed in later chambers; walls thick, usually tan, ornamented with eight longitudinal, evenly spaced prominent costae that diminish in height at apertural end and coalesce at apical end to form a stout spine, aperture somewhat produced, radiate.

Measurements. Length 2.52 mm, maximum diameter 0.64 mm.

Distribution. Moritzian through Brunswickian Stages.

Locality. Sample F-158, Santa María River, Tierra del Fuego.

Depository. Holotype: Museo Nacional de Historia Natural, Santiago, Chile, cat. no. SGO..p.m. Pi. 162. Paratype: U.S. National Museum, Washington, D.C., cat. no. 688432.

Family POLYMORPHINIDAE d'Orbigny, 1839
Subfamily POLYMORPHININAE d'Orbigny, 1839

Genus *Polymorphina* d'Orbigny, 1839
Polymorphina martinezi Cañon and Ernst, n. sp.
(Pl. 2, figs. 8a, b)

Description. Test elongate, slightly twisted, compressed, light tan to dark reddish-brown, biserial; chambers indistinct to distinct and slightly inflated; sutures curve downward from median line and are slightly depressed, especially in later chambers; aperture terminal, radiate, produced on some specimens.

Measurements. Length 1.20 mm, width 0.60 mm.

Remarks. This species forms an excellent zone over a wide area.

Distribution. Esperanzian and Rinconian Stages.

Locality. ENAP Victoria Norte no. 2, core no. 1, 2,250–2,253.6 m.

Depository. Holotype: Museo Nacional de Historia Natural, Santiago, Chile, cat. no. SGO. p.m. Pi. 156. Paratype: U.S. National Museum, Washington, D.C., cat. no. 688433.

Superfamily BULIMINACEA Jones, 1875
Family SPHAEROIDINIDAE Cushman, 1927

Genus *Sphaeroidina* d'Orbigny, 1826
Sphaeroidina bulloides d'Orbigny
(Pl. 2, figs. 9a, b)

Sphaeroidina bulloides d'Orbigny, 1826. Loeblich and Tappan, 1964, p. C547, Fig. 423, 1–3.

Remarks. This species is very widespread, both areally and vertically in the geologic column. It is generally found living in ocean depths greater than 700 m.

Distribution. Lower Gaviotian to upper Lazian Stages. Most abundant in upper Miradorian.

Locality. Sample F-315, Puerto Nuevo, northeast coast Bahía Inútil, Tierra del Fuego.

Family BOLIVINITIDAE Cushman, 1927

Genus *Bolivina* d'Orbigny, 1839
Bolivina incrassata Reuss
(Pl. 2, figs. 10a, b)

Bolivina incrassata Reuss, 1851. Cushman, 1927, p. 86, Pl. 12, figs. 1a, b.

Remarks. The Magallanes specimens closely resemble forms from the south-central United States which were figured by Cushman. This species is widely distributed and well represented in the Cretaceous and Paleocene of the Magallanes Basin. In comparison with forms figured from other localities, our specimens are not as wide in relation to height and their sutures bend more downward.

Distribution. Germanian and Riescoian Stages.

Locality. Sample MX-2.022, Rocallosa Point, Riesco Island.

Bolivina incrassata Reuss var. *gigantea* Wicher
(Pl. 2, figs. 11a, b)

Bolivina incrassata Reuss var. *gigantea* Wicher, 1949, p. 57 (Serbian), p. 85 (English), Pl. 5, figs. 2, 3.

Remarks. Our specimens also resemble *B. incrassata* Reuss var. *crassa* Vasilenko and Myatliuk from the upper Maestrichtian of Russia and *B. incrassata* Reuss var. *lata* Egger from Germany.

Distribution. Upper part Riescoian Stage.

Locality. Sample MX-2.017, Rocallosa Point, Riesco Island.

Family BULIMINIDAE Jones, 1875
Subfamily BULIMININAE Jones, 1875

Genus *Bulimina* d'Orbigny, 1826
Bulimina gonzalezi Cañon and Ernst, n. sp.
(Pl. 2, figs. 12a, b)

Description. Test triserial, spined initial end; chambers distinct in later part; sutures slightly depressed in last two whorls; aperture extends up from near base of last chamber to terminal end.

Measurements. Length 0.35 mm, maximum diameter 0.15 mm.

Remarks. This form is similar to *B. versa* Cushman and Parker but differs from it by having an apical spine, smooth walls without costae, and a more twisted chamber arrangement. It also resembles *B. reussi* Morrow var. *navarroensis* Cushman and Parker.

Distribution. Oazian and Germanian Stages.

Locality. Sample JC-248, Canelos River, south coast Seno Otway, Brunswick Península.

Depository. Holotype: Museo Nacional de Historia Natural, Santiago, Chile, cat. no. SGO. p.m. Pi. 164. Paratype: U.S. National Museum, Washington, D.C., cat. no. 688434.

Genus *Praeglobobulimina* Hofker, 1951
Praeglobobulimina kickapooensis (Cole)
(Pl. 2, figs. 13a-c)

Bulimina kickapooensis Cole, 1938, p. 45, Pl. 3, fig. 5.

Remarks. This species is widely distributed in the Magallanes Basin and appears to be the same as the forms figured from elsewhere in the world.

Distribution. Marks top of Germanian Stage and continues into Lazian Stage.

Locality. Sample V-9, Chorrillo García, southeast coast Brunswick Península.

Praeglobobulimina pupoides (d'Orbigny)
(Pl. 2, figs. 14a, b)

Bulimina pupoides d'Orbigny, 1846. Todd and Kniker, 1952, p. 19, Pl. 4, figs. 1, 2.

Distribution. Gaviotian through Lazian Stages. Most abundant in upper part of Miradorian Stage.

Locality. Sample Ruby 4, south coast Seno Otway, Brunswick Península.

Family UVIGERINIDAE Haeckel, 1894

Genus *Rectuvigerina* Mathews, 1945
Rectuvigerina ongleyi (Finlay)
(Pl. 2, figs. 15a, b)

Siphogenerina ongleyi Finlay, 1939, p. 111, Pl. 13, figs. 42, 43.

Remarks. Like most of the Uvigerinidae, this species varies considerably in shape and ornamentation with stratigraphic position. In the Magallanes Basin, this species in a downward stratigraphic direction becomes more inflated with finer costae, especially in the later chambers.

Distribution. Rosarian Stage; first occurrence marks top of stage.

Locality. Sample F-234, north coast Bahía Inútil, Discordia River, Tierra del Fuego.

Genus *Trifarina* Cushman, 1923
Trifarina angulosa (Williamson)
(Pl. 2, figs. 16a, b)

Uvigerina angulosa Williamson, 1858.
Trifarina angulosa (Williamson) Loeblich and Tappan, 1964, p. C571, Fig. 450, 1-3.

Distribution. MacPhearsonian through upper Miradorian Stages. Most abundant in upper Miradorian Stage.
Locality. Sample F-539, south coast Brush Lake, Tierra del Fuego.

Superfamily DISCORBACEA Ehrenberg, 1838
Family DISCORBIDAE Ehrenberg, 1838
Subfamily DISCORBINAE Ehrenberg, 1838

Genus *Discorbis* Lamarck, 1804
Discorbis minima Vieaux
(Pl. 2, figs. 17a-c)

Discorbis minima Vieaux, 1941, p. 627, Pl. 85, figs. 10a-c.

Remarks. Our specimens are very similar to the figured Lower Cretaceous form from the Denton Formation of Texas but differ from it by being more compressed. The Magallanes Basin forms have walls that appear to be more roughened by a surplus of calcareous material; therefore, they may represent a new species.
Distribution. Lower Lazian to upper part of Tenerifian Stage.
Locality. ENAP Mellizos no. 1, core no. 1, 1,789-1,795 m.

Genus *Buccella* Andersen, 1952
Buccella depressa Andersen
(Pl. 3, figs. 1a-c)

Buccella depressa Andersen, 1952, p. 145.

Remarks. Our specimens compare very well with the Holocene forms figured by Andersen from the Falkland Islands. Some of our specimens are also very much like *B. parkerae*, a Miocene species found in the Temblor Formation of California.
Distribution. MacPhearsonian through Gaviotian Stages with greatest abundance in the MacPhearsonian Stage.
Locality. ENAP Paraguaya no. 1, 401-405.5 m.

Genus *Epistominella* Husezima and Maruhasi, 1944
Epistominella texana (Cushman)
(Pl. 3, figs. 2a-c)

Pulvinulinella texana Cushman, 1938b, p. 49, Pl. 8, figs. 8a-c.

Remarks. Our specimens are very similar to the form figured by Cushman from near the base of the upper part of the Taylor marl of Cretaceous age.
Distribution. Oazian through Riescoian Stages.
Locality. Sample JC-354, south of Prat Point, south coast Seno Otway.

Superfamily ROTALIACEA Ehrenberg, 1839
Family ELPHIDIIDAE Galloway, 1933
Subfamily ELPHIDINAE Galloway, 1933

Genus *Elphidium* de Montfort, 1808
Elphidium aguafrescaense Todd and Kniker
(Pl. 3, figs. 3a, b)

Elphidium aguafrescaense Todd and Kniker, 1952, p. 19, Pl. 3, figs. 36a, b.

Distribution. Manzanian Stage. Abundant in upper part.
Locality. T-69, Cerro Laurita, Brunswick Península.

Elphidium patagonicum Todd and Kniker
(Pl. 3, figs. 4a, b)

Elphidium patagonicum Todd and Kniker, 1952, p. 18, Pl. 3, figs. 35a, b.

Remarks. This species is about twice as broad as *E. aguafrescaense* and is persistently rare in the upper part of the Brunswickian Stage, becoming more abundant in the lower part. A varietal form *E.* cf. *patagonicum* differs from the typical form by being smaller, more compressed, rounder in cross section and having more uniform retral processes.
Distribution. Moritzian through Brunswickian Stages.
Locality. ENAP well P-7, 118-121 m.

Elphidium skyringense Todd and Kniker
(Pl. 3, figs. 5a, b)

Elphidium skyringense Todd and Kniker, 1952, p. 18, Pl. 3, figs. 39a, b.

Remarks. This species is easily distinguished by its outwardly spiraling retral processes.
Distribution. Restricted to upper part of Manzanian Stage where it is abundant.
Locality. T-211, Agua Fresca River, Brunswick Península.

Genus *Cribroelphidium* Cushman and Brönnimann, 1948
Cribroelphidium cf. *strattoni* (Applin)
(Pl. 3, figs. 6a, b)

Polystomella strattoni Applin and others, 1925, p. 100, Pl. 3, figs. 9, 10.

Remarks. Our specimens differ from *C. strattoni* by being broader with a rounder periphery. Otherwise, they appear to be the same species.
Distribution. Rosarian and Cameronian Stages.
Locality. ENAP San Antonio no. 1, core no. 178, 1,503-1,506 m.

Superfamily GLOBIGERINACEA Carpenter, Parker and Jones, 1862
Family HETEROHELICIDAE Cushman, 1927
Subfamily HETEROHELICINAE Cushman, 1927

Genus *Heterohelix* Ehrenberg, 1843
Heterohelix globulosa (Ehrenberg)
(Pl. 3, figs. 7a, b)

Textularia globulosa Ehrenberg, 1840.
Heterohelix globulosa (Ehrenberg) Loeblich and Tappan, 1964, p. C652, Fig. 523, 5-7.

Distribution. Marks top of Riescoian Stage; abundant in Lazian Stage.
Locality. ENAP San Sebastián no. 1, 1,661-1,670 m.

Heterohelix moremani (Cushman)
(Pl. 3, figs. 8a, b)

Gümbelina moremani Cushman, 1938a, p. 10, Pl. 2, figs. 1-3.

Remarks. This species differs from *H. globulosa* (Ehrenberg) by generally having more chambers and a more slender and less tapering test. It was originally described from the Cretaceous Eagle Ford shale near Itasca, Texas.
Distribution. Peninsulian Stage, index species.
Locality. ENAP Estancia Nueva no. 1, core no. 1, 1,977.5-1,983.5 m.

Family ROTALIPORIDAE Sigal, 1958
Subfamily HEDBERGELLINAE Loeblich and Tappan, 1961

Genus *Hedbergella* Brönnimann and Brown, 1958
Hedbergella planispira (Tappan)
(Pl. 3, figs. 9a, b)

Globigerina planispira Tappan, 1940, p. 122, Pl. 19, fig. 12.

Remarks. This species is smaller than *Globigerina* cf. *cretacea* d'Orbigny and has more small bulbous chambers in each whorl.
Distribution. Peninsulian Stage, index species.
Locality. ENAP Estancia Nueva no. 1, core no. 1, 1,977.5-1,983.5 m.

Family GLOBOTRUNCANIDAE Brotzen, 1942

Genus *Globotruncana* Cushman, 1927
Globotruncana chileana Cañon and Ernst, n. sp.
(Pl. 3, figs. 10a-c)

Description. Test free, trochospiral, biconvex, compressed; periphery slightly lobate and weakly bicarinate, carinal band narrow and slightly inclined; chambers evolute, slowly increasing in size especially in the last whorl which has seven chambers, dorsal chambers petaloid; dorsal sutures curved, raised, weakly beaded in the early chambers, ventral sutures radial, depressed; walls smooth to rugose; umbilicus depressed, circular, covered with slightly beaded wall material; aperture indistinct, probably umbilical.

Measurements. Maximum diameter 0.42 mm, thickness 0.17 mm.

Remarks. This species resembles *G. aspera* but differs from it by not having well-developed raised sutures and keels. *G. (G.) bulloides* Vogler subsp. *naussi* Gandolfi is more lobate than *G. chileana*.

Distribution. Lazian Stage, very abundant.

Locality. ENAP San Sebastián no. 1, 1,622-1,631 m.

Depository. Holotype: Museo Nacional de Historia Natural, Santiago, Chile, cat. no. SGO. p.m. Pi. 155. Paratype: U.S. National Museum, Washington, D.C., cat. no. 688435.

Globotruncana (Globotruncana) marginata (Reuss)
(Pl. 3, figs. 11a-c)

Rosalina marginata Reuss, 1845.
Globotruncana (Globotruncana) marginata (Reuss) Edgell, 1957, p. 114, Pl. 2, figs. 4-6.

Remarks. Identification of our specimens is uncertain because their average keel development is not as pronounced and their tests are not as compressed as those of specimens described from other localities.

Distribution. Lazian Stage.

Locality. ENAP San Sebastián no. 1, 1,622-1,631 m.

Globotruncana (Globotruncana) lapparenti Brotzen
cf. subsp. *tricarinata* (Quereau)
(Pl. 3, figs. 12a-c)

Pulvinulina tricarinata Quereau, 1893.
Globotruncana (Globotruncana) lapparenti Brotzen cf. subsp. *tricarinata* (Quereau) Edgell, 1957, p. 113, Pl. 3, figs. 1-3.

Remarks. The chamber arrangement and shape of our specimens compare well with Australian forms figured by Edgell. Ours differ by having more poorly developed keels and beaded sutures.

Distribution. Lazian and Peninsulian Stages, index species.

Locality. ENAP San Sebastián no. 1, 1,622-1,631 m.

Family HANTKENINIDAE Cushman, 1927
Subfamily HASTIGERININAE Bolli, Loeblich and Tappan, 1957

Genus *Hastigerina* Thomson, 1876
Hastigerina escheri escheri (Kaufmann)
(Pl. 4, figs. 1a, b)

Nonionina escheri Kaufmann, 1875.
Globigerinella escheri escheri (Kaufmann), Brönnimann, 1952, p. 46, Figs. 22, 23.

Remarks. Loeblich and Tappan have reassigned *Globigerinella* to *Hastigerina*, and we have followed their system. Although our specimens resemble the *Globanomalina* forms figured by Loeblich and Tappan (1964, p. C664, Fig. 531, 5-8), they do not have the apertural lip characteristic of this genus. *H. escheri escheri* is common to abundant in the *Globotruncana lapparenti* s. l. zone of Trinidad (Turonian-Senonian). It was originally reported from the Upper Cretaceous Seewerkalk of Switzerland (top of Cenomanian and Turonian-Senonian) and from the white chalk of England.

Distribution. Peninsulian Stage, index species.

Locality. ENAP Estancia Nueva no. 1, core no. 1, 1,977.5-1,983.5 m.

Hastigerina iota (Finlay)
(Pl. 4, figs. 2a, b)

Nonion iota **Finlay, 1940, p. 456, Pl. 65, fig. 109, holotype.**

Remarks. The generic classification of this species has been changed from *Nonion* to *Hastigerina* because it has a much broader interiomarginal aperture than *Nonion*. In New Zealand, this species ranges from lower Bartonian (middle Eocene) to Kaiatan (lower Oligocene).

Distribution. Lower part of Brunswickian Stage, index species.

Locality. ENAP Manzano no. 7, 3,285-3,294 m.

Family GLOBOROTALIIDAE Cushman, 1927
Subfamily GLOBOROTALIINAE Cushman, 1927

Genus *Globorotalia* Cushman, 1927
Globorotalia cf. *crassata* (Cushman) var.
aequa Cushman and Renz

Globorotalia crassata **(Cushman) var.** *aequa* **Cushman and Renz, 1942, p. 12, Pl. 3, figs. 3a-c.**

Remarks. This variety from the Eocene of Trinidad differs from our specimens by having smoother walls and a rounder periphery.

Distribution. Lower part of Brunswickian Stage, index species.

Locality. ENAP Manzano no. 7, 3,285-3,294 m.

Family GLOBIGERINIDAE Carpenter, Parker and Jones, 1862
Subfamily GLOBIGERININAE Carpenter, Parker and Jones, 1862

Genus *Globigerina* d'Orbigny, 1826
Globigerina cretacea d'Orbigny
(Pl. 4, figs. 3a-c)

Globigerina cretacea **d'Orbigny, 1840. Morrow, 1934, p. 198, Pl. 30, figs. 7, 8, 10a, b.**

Remarks. This species is widespread in the Magallanes Basin.
Distribution. Less abundant in Riescoian Stage, abundant in Lazian Stage. Occurrences above Riescoian are presumed to be out of place.
Locality. ENAP San Sebastián no. 1, 1,631-1,677 m.

Globigerina triloculinoides Plummer
(Pl. 4, figs. 4a, b)

Globigerina triloculinoides Plummer, 1926, p. 134-135, Pl. 8, fig. 10.

Remarks. This species occurs frequently throughout the Magallanes Basin sections, but occurrences in stages other than the Brunswickian and Manzanian may result from ditch sample contamination.
Distribution. Brunswickian and Manzanian Stages.
Locality. ENAP Manzano no. 7, core no. 1, 2,775-2,780 m.

Globigerina wenzeli Cañon and Ernst, n. sp.
(Pl. 4, figs. 5a-c)

Description. Test free, a high trochospiral coil with about three whorls; chambers spherical to subglobular, usually six in each whorl; sutures depressed; walls calcareous, finely perforated, rather smooth; aperture small, at the base of the last chamber, opening into a deep umbilical pit.
Measurements. Maximum diameter 0.43 mm, thickness 0.35 mm.
Remarks. This species is closely related to *Globigerina paradubia* Sigal. It also has a chamber arrangement similar to *Globoquadrina altispira globosa* Bolli but differs by having a simple arched opening rather than an aperture with serrated flaps.
Distribution. Lazian and Peninsulian Stages.
Locality. ENAP Espora no. 1, 1,427-1,475 m.
Depository. Holotype: Museo Nacional de Historia Natural, Santiago, Chile, cat. no. SGO. p.m. Pi. 159. Paratype: U.S. National Museum, Washington, D.C., cat. no. 688436.

Subfamily ORBULININAE Schultze, 1854

Genus *Candeina* d'Orbigny, 1839
Candeina cecionii Cañon and Ernst, n. sp.
(Pl. 4, figs. 6a-c)

Description. Test free, trochoid, low spired, quadrate in side view; chambers inflated, four in last whorl; sutures depressed; series of small rounded apertures along sutures in adult.
Measurements. Maximum diameter 0.20 mm, thickness 0.16 mm.
Remarks. The quadrate shape of this species sets it apart from other *Candeina* species, which are generally trigonal in form.
Distribution. Lower Gaviotian and Miradorian Stages. Older occurrences are presumed to be contaminants.

Locality. Sample S-13, south coast Seno Otway, Brunswick Península.
Depository. Holotype: Museo Nacional de Historia Natural, Santiago, Chile, cat. no. SGO. p.m. Pi. 163. Paratype: U.S. National Museum, Washington, D.C., cat. no. 688437.

Superfamily ORBITOIDACEA Schwager, 1876
Family CIBIDICIDAE Cushman, 1927
Subfamily PLANULININAE Bermudez, 1952

Genus *Planulina* d'Orbigny, 1826
Planulina popenoei (Trujillo)
(Pl. 4, figs. 7a-c)

Anomalina popenoei Trujillo, 1960, p. 335, Pl. 48, figs. 9a-c.

Remarks. Specimens from the Magallanes Basin are nearly identical to those reported by Trujillo from the Upper Creteceous shales of the San Joaquin Valley, California.
Distribution. Lazian Stage, index species.
Locality. ENAP San Sebastián no. 1, 1,616-1,625 m.

Subfamily CIBICIDINAE Cushman, 1927

Genus *Cibicides* de Montfort, 1808
Cibicides cf. *djaffaensis* Sigal
(Pl. 4, figs. 8a-c)

Cibicides cf. *djaffaensis* Sigal, 1952, p. 14, Fig. 5.

Remarks. This species is characterized principally by its bi-umbonate profile.
Distribution. Upper part of Tenerifian Stage.
Locality. ENAP Monte Aymond no. 2, core no. 1, 2,009.5-2,015 m.

Family CASSIDULINIDAE d'Orbigny, 1839

Genus *Cassidulina* d'Orbigny, 1826
Cassidulina cf. *brocha* Poag
(Pl. 4, figs. 10a, b)

Cassidulina brocha Poag, 1966, p. 426, Pl. 8, figs. 33-35.

Remarks. Our specimens are very similar to *C. brocha* but differ by having a rounder periphery, straighter sutures, and a more elongate aperture extending nearly to the periphery. In many respects they also closely resemble *C. rotulita* Poag figured on the same plate as *C. brocha.*
Distribution. Occurs in the lower part of the Gaviotian Stage and in the Miradorian Stage. Sporadic lower occurrences are presumed to be contaminants.
Locality. ENAP Manatiales no. 1, 762-765.1 m.

Superfamily CASSIDULINACEA d'Orbigny, 1839
Family CAUCASINIDAE N. K. Bykova, 1959
Subfamily FURSENKOININAE Loeblich and Tappan, 1961

Genus *Virgulinella* Cushman, 1932
Virgulinella severini Cañon and Ernst, n. sp.
(Pl. 4, figs. 9a, b)

Description. Test free, elongate, round in section with spine at proximal end, early stage triserial, later biserial; chambers becoming progressively more inflated toward apertural end, later chambers with small arched opening along the sutures; walls calcareous, finely perforated, smooth, semitranslucent; aperture a terminal loop, open, extending down to the suture, very small, pointed tooth.
Measurements. Length 1.10 mm, maximum diameter 0.30 mm.
Remarks. The presence of arched openings along the sutures is variable and is usually present only in the last few chambers.
Distribution. Cameronian and Moritzian Stages. Occurrences below Moritzian are presumed to be contaminants.
Locality. Sample F-173, Discordia River, north coast Bahía Inútil, Tierra del Fuego.
Depository. Holotype: Museo Nacional de Historia Natural, Santiago, Chile, cat. no. SGO. p.m. Pi. 157. Paratype: U.S. National Museum, Washington, D.C., cat. no. 688438.

Family NONIONIDAE Schultze, 1854
Subfamily CHILOSTOMELLINAE Brady, 1881

Genus *Allomorphina* Reuss, 1849
Allomorphina conica Cushman and Todd
(Pl. 4, figs. 11a, b)

Allomorphina conica Cushman and Todd, 1949, p. 62, Pl. 11, figs. 8a-c.

Remarks. Our specimens closely resemble this species, first described from the Cretaceous Lizard Springs formation of Trinidad.
Distribution. Riescoian through Peninsulian Stages.
Locality. ENAP Manzano no. 7, 3,465-3,474 m.

Subfamily NONIONINAE Schultze, 1854

Genus *Florilus* de Montfort, 1808
Florilus cf. *boueanus* (d'Orbigny)
(Pl. 4, figs. 12a, b)

Nonionina boueana d'Orbigny, 1846.
Nonion boueanum (d'Orbigny) Marks, 1951, p. 48, Pl. 5, figs. 17a, b.

Remarks. The Magallanes Basin specimen closely resembles that figured by Marks from the Vienna Basin.

Distribution. Lower Gaviotian and upper Miradorian Stages.
Locality. Sample F-538, south coast of Brush Lake, Tierra del Fuego.

Florilus scaphus (Fichtel and Moll)
(Pl. 4, figs. 13a, b)

Nautilus scapha Fichtel and Moll, 1798.
Nonion scaphum (Fichtel and Moll) Cushman, 1930, p. 5, Pl. 2, figs. 3, 4.

Remarks. This species is quite abundant in shallow-water upper Tertiary sediments of the Magallanes Basin.
Distribution. Gaviotian through upper Miradorian Stages. Lower occurrences are presumed to be contaminants.
Locality. Sample F-538, south coast of Brush Lake, Tierra del Fuego.

Genus *Nonionella* Cushman, 1926
Nonionella auris (d'Orbigny)
(Pl. 4, figs. 14a-c)

Valvulina auris d'Orbigny, 1839.
Nonionella auris (d'Orbigny) Cushman, 1939, p. 33, Pl. 9, fig. 4.

Remarks. The Magallanes Basin forms also resemble *N. pseudo-auris* Cole, *N. modesta* Galloway and Heminway, and *N. pauciloba* Cushman.
Distribution. Miradorian through Cameronian Stages. Lower occurrences are presumed to be contaminants.
Locality. Sample F-538, south coast of Brush Lake, Tierra del Fuego.

Nonionella pulchella Hada
(Pl. 4, figs. 15a-c)

Nonionella pulchella Hada, 1931, p. 120, Pl. 121, figs. 79a-c.

Remarks. The final chamber of the Chilean form is wider and rounder than the specimen from Japan figured by Hada. In other respects, it seems sufficiently similar to be referred to this species. It also resembles *N. miocenica* var. *stella* Cushman and Moyer but differs by having ten chambers per whorl rather than eight.
Distribution. MacPhearsonian Stage.
Locality. ENAP Catalina no. 1, 172-176 m.

Genus *Pullenia* Parker and Jones, 1862
Pullenia bulloides (d'Orbigny)
(Pl. 5, figs. 1a, b)

Nonionina bulloides d'Orbigny, 1846.
Pullenia bulloides (d'Orbigny) Cushman and Todd, 1943, p. 13, Pl. 2, figs. 15-18.

Remarks. Typical representatives of this species range from Miradorian through Moritzian Stages and are most abundant in the Cameronian Stage. The first persistent occurrence of this species usually marks the top of the Miradorian Stage.

Distribution. Miradorian through upper Lazian Stages. Lazian occurrences may be contaminants.
Locality. Sample F-311, Bahía Inútil, Tierra del Fuego.

Pullenia natlandi Cañon and Ernst, n. sp.
(Pl. 5, figs. 2a, b)

Description. Test compressed, usually planispiral with some specimens slightly asymmetric in transverse section, involute; chambers subglobular, usually four in last whorl although a few specimens have five; sutures distinct, depressed, radial forming right angle crosses on sides; walls calcareous, smooth, finely perforated; umbilici depressed; aperture a narrow interiomarginal slit extending from umbilicus to umbilicus.
Measurements. Length 0.36 mm, width 0.28 mm.
Remarks. This species differs from other figured *Pullenia* sp. by having four compressed chambers in the last whorl with the front face of the last chamber about twice as high as that of other *Pullenia* sp. It resembles *P. cretacea* (Cushman) but differs by having four to five chambers in the last whorl instead of five to six and by having a very inflated last chamber.
Distribution. Upper part of Tenerifian Stage.
Locality. ENAP Posesión no. 1, 1,449-1,452 m.
Depository. Holotype: Museo Nacional de Historia Natural, Santiago, Chile, cat. no. SGO. p.m. Pi. 160. Paratype: U.S. National Museum, Washington, D.C., cat. no. 688439.

Family ALABAMINIDAE Hofker, 1951

Genus *Gyroidina* d'Orbigny, 1926
Gyroidina infrafosa Finlay
(Pl. 5, figs. 3a-c)

Gyroidina infrafosa Finlay, 1940, p. 462, Pl. 66, figs. 181-183.

Remarks. The rugose walls of this species suggest that it might better be placed in *Anomalina* rather than *Gyroidina* which, as a rule, has rather smooth walls.
Distribution. Oazian through Esperanzian Stages.
Locality. Sample G-399, Rocallosa Point, Riesco Island.

Gyroidina soldanii d'Orbigny
(Pl. 5, figs. 4a-c)

Gyroidina soldanii d'Orbigny, 1826. Todd and Kniker, 1952, p. 24, Pl. 4, figs. 20a-c.

Remarks. Typical representatives of this species are most abundant in the Cameronian Stage.
Distribution. Indigenous from Gaviotian through Cameronian Stages. Lower occurrences are presumed to be contaminants.
Locality. Sample F-293, north coast Bahía Inútil, Tierra del Fuego.

Family OSANGULARIIDAE Loeblich and Tappan, 1964

Genus *Osangularia* Brotzen, 1940
Osangularia brunswickensis Todd and Kniker
(Pl. 5, figs. 5a-c)

Osangularia brunswickensis Todd and Kniker, 1952, p. 24, Pl. 4, figs. 22a-c.

Remarks. This species is similar to *O. plummerae* Brotzen and *O. lens* Brotzen. It differs from *O. tenuis carinata* (Cushman and Siegfus) by having a larger and more compressed test.
Distribution. Manzanian Stage.
Locality. ENAP well P-7, 718-723 m.

Family ANOMALINIDAE Cushman, 1927
Subfamily ANOMALININAE Cushman, 1927

Genus *Anomalina* d'Orbigny, 1826
Anomalina rubiginosa Cushman
(Pl. 5, figs. 6a-c)

Anomalina rubiginosa Cushman, 1926, p. 607, Pl. 21, figs. 6a-c.

Remarks. Our specimens are similar in general shape to those figured by Cushman from the Velasco shale of the Tampico Embayment, Mexico. They differ mainly by not being as reticulate on the ventral side.
Distribution. From Moritzian Stage into Riescoian Stage.
Locality. Sample G-47, Bahía Fuentes, Brunswick Península.

Genus *Cibicidoides* Thalmann, 1939
Cibicidoides Semiumbilicatus Toutkovski
(Pl. 5, figs. 7a-c)

Discorbina semiumbilicata Toutkovski, 1887, p. 358, Pl. 5, fig. 5.

Distribution. Lazian Stage, index species.
Locality. ENAP San Sebastián no. 1, 1,616-1,625 m.

Superfamily ROBERTINACEA Reuss, 1850
Family CERATOBULIMINIDAE Cushman, 1927
Subfamily EPISTOMININAE Wedekind, 1937

Genus *Hoeglundina* Brotzen, 1948
Hoeglundina elegans (d'Orbigny)
(Pl. 5, figs. 8a-c)

Rotalia (Turbinulina) elegans d'Orbigny, 1826.
Hoeglundina elegans (d'Orbigny) Loeblich and Tappan, 1964, p. C775, Fig. 636, 3a-c.

Remarks. This species is very similar to *H. eocenica* (Cushman and Hanna). The main difference between them is that *eocenica* has five to seven chambers in the last whorl instead of seven to nine as in *elegans*.

Distribution. Lower Gaviotian to upper Lazian Stage. First abundant occurrence marks top of Miradorian Stage.

Locality. Sample F-313, Bahía Inútil Hotel, Tierra del Fuego.

Hoeglundina porcellanea (Brückmann)
(Pl. 6, 1a-c)

Epistomina porcellanea Brückmann, 1904, p. 26, Pl. 4, figs. 17-19.

Distribution. Rinconian Stage.
Locality. ENAP Ciaike no. 1, core no. 4, 3,260 m.

Genus *Reinholdella* Brotzen, 1948
Reinholdella fuenzalidai Cañon and Ernst, n. sp.
(Pl. 6, figs. 2a-c)

Description. Test free, trochoid, plano-convex; five chambers per whorl, three to four whorls on convex dorsal side; sutures distinct, depressed, very oblique dorsally and radial ventrally; walls calcareous, smooth, finely perforate; aperture a low interiomarginal arch near periphery on ventral side; previous apertures covered with supplementary plates.

Measurements. Maximum diameter 0.33 mm, thickness 0.17 mm.

Remarks. The only specimens recovered to date are in the form of pyrite casts with the original wall material dissolved so that generic wall characteristics must be assumed.

Distribution. Rinconian Stage.
Locality. ENAP Ciaike no. 1, core no. 4, 3,261 m.
Depository. Holotype: Museo Nacional de Historia Natural, Santiago, Chile, cat. no. SGO. p.m. Pi. 161. Paratype: U.S. National Museum, Washington, D.C., cat. no. 688440.

Reinholdella cf. *quadrilocula* Subbotina and Datta
(Pl. 6, figs. 3a-c)

Reinholdella cf. *quadrilocula* Subbotina, Datta and Srivastava, 1960, p. 39, Pl. 3, figs. 8a-c.

Remarks. Our specimens of this species recovered to date are in the form of pyritized casts, and definite identification is difficult. However, our forms very closely resemble that described by Subbotina and others (1960).

Distribution. Rinconian Stage.
Locality. ENAP Ciaike no. 1, core no. 4, 3,262 m.

Family ROBERTINIDAE Reuss, 1850

Genus *Robertina* d'Orbigny, 1846
Robertina arctica d'Orbigny
(Pl. 6, figs. 4a–c)

Robertina arctica d'Orbigny, 1846. Cushman and Parker, 1947, p. 74, Pl. 18, figs. 12a, b.

Distribution. Gaviotian Stage.
Locality. ENAP Rio del Oro no. 1, 595–598 m.

Radiolaria spumellaria? sp. 4
(Pl. 6, figs. 5a, b)

Description. Form is similar morphologically to Radiolaria *Spumellaria?* sp. 1, but differs by being less compressed and twice as large; color ranges from gray to white to pink.
Measurements. Average diameter 0.38 mm.
Remarks. Since it also lacks typical radiolarian structure, it may be something else, but it, too, is good for correlation purposes. It is confined to the cretadura Margas Formation.
Distribution. Tenerifian Stage.
Locality. ENAP Pampa Larga no. 1A, 2,550–2,553 m.

Radiolaria spumellaria? sp. 1
(Pl. 6, figs. 6a, b)

Description. Form is round in side view and oval in cross section; lacks typical radiolarian structure; composed of siliceous material (does not effervesce in HCl); color translucent to opaque white.
Measurements. Average diameter 0.38 mm.
Remarks. Because this form does not exhibit typical radiolarian structure, it may belong to some other faunal group or may be some type of nonbiogenic siliceous pellet. Regardless of its origin, it is important because it is good for correlation purposes.
Distribution. Highest occurrences at top of Miradorian Stage. Common through Cameronian and very abundant through Moritzian and Tenerifian Stages.
Locality. ENAP Manzano no. 7, 2,481–2,490 m.

Radiolaria spumellaria? sp. 2

Description. This form is round in side view and flattened in edge view with nearly parallel sides; periphery rounded; walls siliceous, usually structureless; color dark-gray to black.
Measurements. Average diameter 0.18 mm.
Remarks. This form may be closely related to Radiolaria, *Spumellaria?* sp. 1. The base of the abundant Radiolaria, *Spumellaria?* sp. 1 zone in the ENAP Punta del Cerro no. 1 is at 2,636 m and the top of the black Radiolaria, *Spumellaria?*

sp. 2 is at 2,483 m. In the overlap between 2,483 and 2,636 m, the white form sp. 1 gives way to the black form sp. 2, which becomes abundant at 2,519 m.

Distribution. The first persistently abundant occurrence at top of Peninsulian Stage.

Locality. ENAP Punta del Cerro no. 1, 2,519-2,528 m.

Radiolaria spumellaria? sp. 5
(Pl. 6, figs. 7a, b)

Description. The general shape of this form may be referred to the genus *Spongasteriscus* which usually has two smaller projections or lobes extending from opposite sides of an elongate central body; color white.

Measurements. Length 0.33 mm, width 0.24 mm, thickness 0.15 mm.

Remarks. Most specimens effervesce in HCl, and this fact tends to place them outside the radiolarian group. It is quite possible that these forms may be internal flattened casts of some calcareous Foraminifera.

Distribution. Abundant in Clarencian Stage with rare scattered occurrences above and below this stage.

Locality. ENAP Manzano no. 7, 2,481-2,490 m.

Radiolaria spumellaria sp. 6
(Pl. 6, fig. 8)

Description. Test spherical; structure typically radiolarian; wall material pyritized; color yellow-brown.

Measurements. Diameter 0.17 mm.

Distribution. Pratian, Esperanzian, and Rinconian Stages.

Locality. ENAP Ciaike no. 1, core no. 4, 3,260-3,262 m.

Radiolaria nassellaria sp. 3
(Pl. 6, 9a, b)

Description. Test elongate, conical, round in cross section; wall material pyritized; ornamented with closely spaced annular ribs, perforate area between ribs.

Measurements. Length 0.41 mm, diameter 0.20 mm.

Distribution. Rinconian Stage.

Locality. ENAP Ciaike no. 1, core no. 4, 3,260-3,262 m.

Acknowledgments

The authors are grateful for the permission given by the officials of Empresa Nacional del Petróleo (ENAP Chile) to publish the ensuing information dealing with the micropaleontology of the Magallanes Basin of southern Chile.

The writers wish to express their appreciation for the help of M. L. Natland in the preparation of the manuscript and plates.

PLATE SECTION

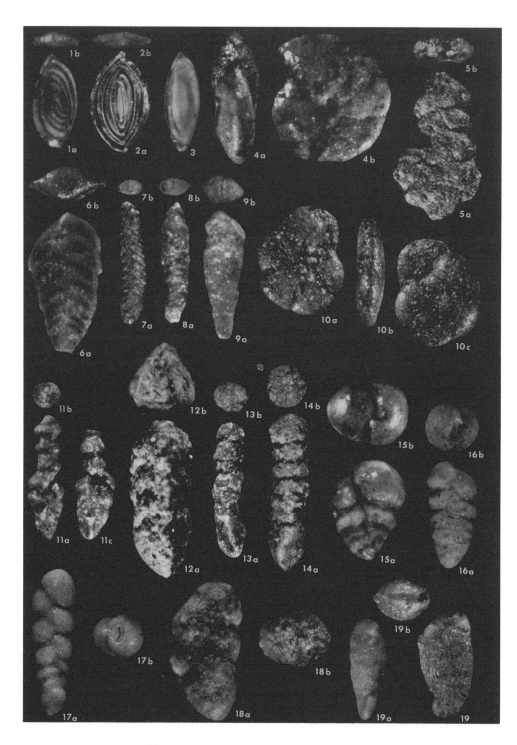

PLATE 1. MAGALLANES BASIN FORAMINIFERA
Geological Society of America Memoir 139

PLATE 1

Figure 1. *Psamminopelta minima* (Cushman and Renz), a, side view; b, apertural view; × 50.

Figure 2, 3. *Psamminopelta venezuelana* (Hedberg); a, side view megalospheric form; b, apertural view megalospheric form; 3, side view microspheric form; × 50.

Figure 4. *Cyclammina cancellata* Brady; a, edge view; b, side view; × 34.

Figure 5. *Ammobaculites barrowensis* Tappan; a, side view; b, apertural view; × 25.

Figure 6. *Spiroplectammina adamsi* Lalicker; a, side view; b, apertural view; × 32.

Figure 7. *Spiroplectammina brunswickensis* Todd and Kniker; a, side view; b, apertural view; × 38.

Figure 8. *Spiroplectammina grzybowskii* Frizzell; a, side view; b, apertural view; × 40.

Figure 9. *Spiroplectammina gutierrezi* Cañon and Ernst, n. sp.; a, side view; b, apertural view; × 35.

Figure 10. *Trochammina* cf. *inflata* (Montagu); a,c, opposite sides; b, edge view; × 45.

Figure 11. *Spiroplectinata annectens* (Parker and Jones); a, side view (short biserial stage); b, apertural view; c, side view (long biserial stage); × 37.

Figure 12. *Tritaxia chileana* (Todd and Kniker); a, side view; b, apertural view; × 29.

Figure 13. *Tritaxia porteri* Cañon and Ernst, n. sp.; a, side view; b, apertural view; × 34.

Figure 14. *Tritaxia rugulosa* (ten Dam and Sigal); a, side view; b, apertural view; × 22.

Figure 15. *Dorothia mordojovichi* Cañon and Ernst, n. sp.; a, side view; b, apertural view; × 50.

Figure 16. *Dorothia principensis* Cushman and Bermúdez; a, side view; b, apertural view; × 43.

Figure 17. *Karreriella cushmani* Finlay; a, side view; b, apertural view; × 30.

Figure 18. *Plectina elongata* Cushman and Bermúdez; a, side view; b, apertural view; × 30.

Figure 19. *Astacolus microdictyotos* Espitalie and Sigal, 1963; a, edge view; d, apertural view; c, side view × 57.

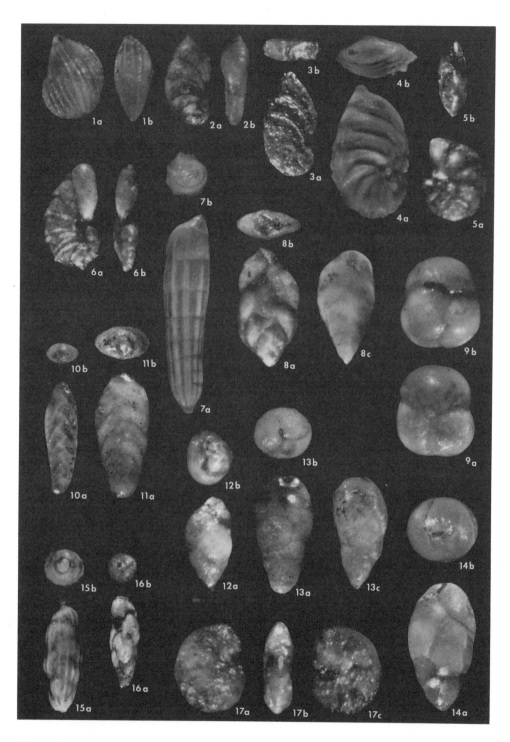

PLATE 2. MAGALLANES BASIN FORAMINIFERA
Geological Society of America Memoir 139

PLATE 2

Figure 1. *Astacolus skyringensis* Todd and Kniker; a, side view; b, edge view; × 46.
Figure 2. *Astacolus stillus* (Terquem); a, side view; b, edge view; × 63.
Figure 3. *Astacolus tricarinellus* (Reuss); a, side view; b, apertural view; × 52.
Figure 4. *Lenticulina* cf. *asperuliformis* (Nuttall); a, side view; b, apertural view; × 32.
Figure 5. *Lenticulina biexcavata* (Myatliuk); a, side view; b, edge view; × 84.
Figure 6. *Lenticulina reyesi* Cañon and Ernst, n. sp.; a, side view; b, edge view; × 47.
Figure 7. *Marginulina knikerae* Cañon and Ernst, n. sp.; a, side view; b, apertural view; × 24.
Figure 8. *Polymorphina martinezi* Cañon and Ernst, n. sp.; a, side view megalospheric form; b, apertural view megalospheric form; c, side view microspheric form; × 45.
Figure 9. *Sphaeroidina bulloides* d'Orbigny; a,b, opposite sides; × 67.
Figure 10. *Bolivina incrassata* Reuss; a, side view; b, apertural view; × 45.
Figure 11. *Bolivina incrassata* Reuss var. *gigantea* Wicher; a, side view; b, apertural view; × 35.
Figure 12. *Bulimina gonzalezi* Cañon and Ernst, n. sp.; a, side view; b, apertural view; × 84.
Figure 13. *Praeglobobulimina kickapooensis* (Cole); a, side view megalospheric form; b, apertural view megalospheric form; c, side view microspheric form; × 35.
Figure 14. *Praeglobobulimina pupoides* (d'Orbigny); a, side view; b, apertural view; × 27.
Figure 15. *Rectuvigerina ongleyi* (Finlay); a, side view; b, apertural view; × 41.
Figure 16. *Trifarina angulosa* (Williamson); a, side view; b, apertural view; × 49.
Figure 17. *Discorbis minima* Vieaux; a,c, opposite sides; b, edge view; × 60.

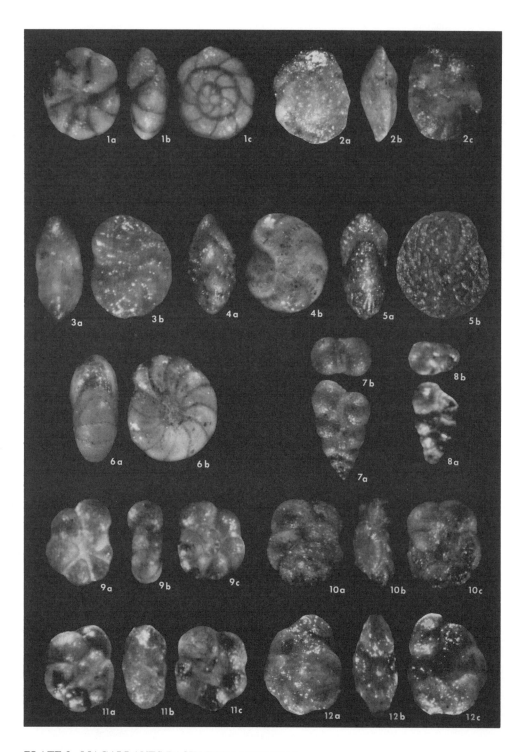

PLATE 3. MAGALLANES BASIN FORAMINIFERA
Geological Society of America Memoir 139

PLATE 3

Figure 1. *Buccella depressa* Andersen; a,c, opposite sides; b, edge view; × 70.
Figure 2. *Epistominella texana* (Cushman); a,c, opposite sides; b, edge view; × 55.
Figure 3. *Elphidium aguafrescaense* Todd and Kniker; a, edge view; b, side view; × 79.
Figure 4. *Elphidium patigonicum* Todd and Kniker; a, edge view; b, side view; × 51.
Figure 5. *Elphidium skyringense* Todd and Kniker; a, edge view; b, side view; × 46.
Figure 6. *Cribroelphidium* cf. *strattoni* (Applin); a, edge view; b, side view; × 58.
Figure 7. *Heterohelix globulosa* (Ehrenberg); a, side view; b, apertural view; × 63.
Figure 8. *Heterohelix moremani* (Cushman); a, side view; b, apertural view; × 120.
Figure 9. *Hedbergella planispira* (Tappan); a,c, opposite sides; b, edge view; × 80.
Figure 10. *Globotruncana chileana* Cañon and Ernst, n. sp.; a,c, opposite sides; b, edge view; × 51.
Figure 11. *Globotruncana (Globotruncana) marginata* (Reuss); a,c, opposite sides; b, edge view; × 133.
Figure 12. *Globotruncana (Globotruncana) lapparenti* Brotzen cf. subsp. *tricarinata* (Quereau); a,c, opposite sides; b, edge view; × 55.

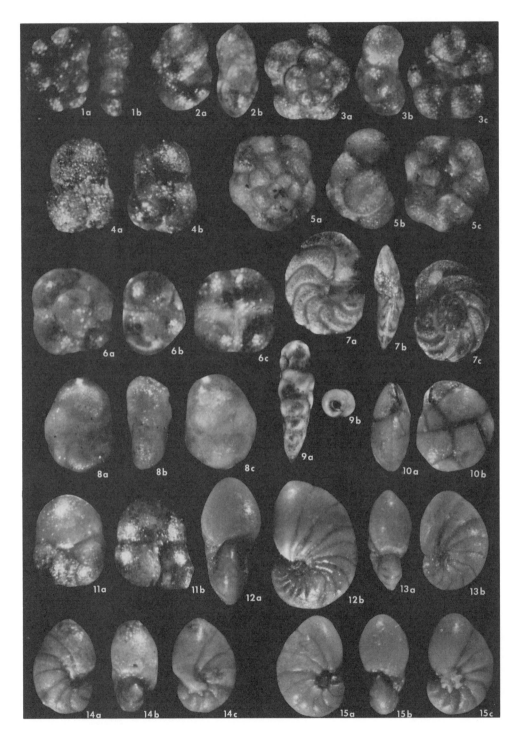

PLATE 4. MAGALLANES BASIN FORAMINIFERA
Geological Society of America Memoir 139

PLATE 4

Figure 1. *Hastigerina escheri escheri* (Kaufmann); a, side view; b, edge view; × 83.
Figure 2. *Hastigerina iota* (Finlay); a, side view; b, edge view; × 169.
Figure 3. *Globigerina cretacea* d'Orbigny; a,c, opposite sides; b, edge view; × 75.
Figure 4. *Globigerina triloculinoides* Plummer; a,b, opposite sides; × 139.
Figure 5. *Globigerina wenzeli* Cañon and Ernst, n. sp.; a,c, opposite sides; b, edge view; × 68.
Figure 6. *Candeina cecionii* Cañon and Ernst, n. sp.; a,c, opposite sides; b, edge view; × 123.
Figure 7. *Planulina popenoei* (Trujillo); a,c, opposite sides; b, edge view; × 54.
Figure 8. *Cibicides* cf. *djaffaensis* Sigal; a,c, opposite sides; b, edge view; × 75.
Figure 9. *Virgulinella severini* Cañon and Ernst, n. sp.; a, side view; b, apertural view; × 34.
Figure 10. *Cassidulina* cf. *brocha* Poag; a, edge view; b, side view; × 49.
Figure 11. *Allomorphina conica* Cushman and Todd; a,b, opposite sides; × 78.
Figure 12. *Florilus* cf. *boueanus* (d'Orbigny); a, edge view; b, side view; × 39.
Figure 13. *Florilus scaphus* (Fichtel and Moll); a, edge view; b, side view; × 57.
Figure 14. *Nonionella auris* (d'Orbigny); a,c, opposite sides; b, edge view; × 76.
Figure 15. *Nonionella pulchella* Hada; a,c, opposite sides; b, edge view; × 75.

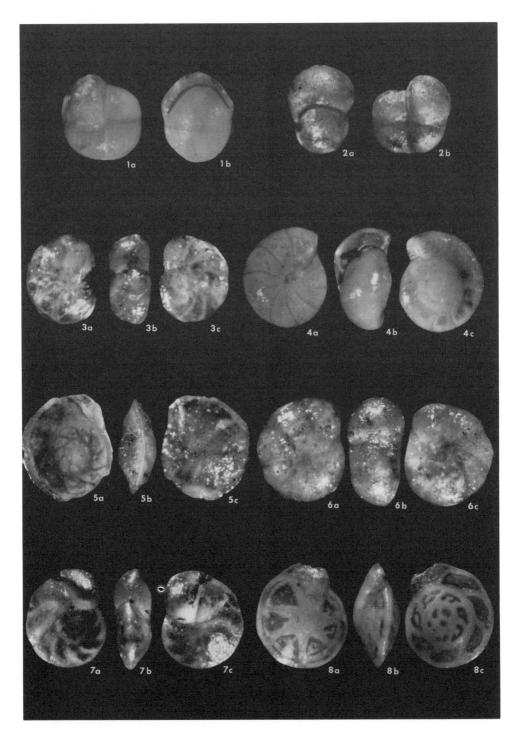

PLATE 5. MAGALLANES BASIN FORAMINIFERA
Geological Society of America Memoir 139

PLATE 5

Figure 1. *Pullenia bulloides* (d'Orbigny); a, side view; b, edge view; × 90.
Figure 2. *Pullenia natlandi* Cañon and Ernst, n. sp.; a, edge view; b, side view; × 62.
Figure 3. *Gyroidina infrafosa* Finlay; a,c, opposite sides; b, edge view; × 67.
Figure 4. *Gyroidina soldanii* d'Orbigny; a,c, opposite sides; b, edge view; × 53.
Figure 5. *Osangularia brunswickensis* Todd and Kniker; a,c, opposite sides; b, edge view; × 45.
Figure 6. *Anomalina rubiginosa* Cushman; a,c, opposite sides; b, edge view; × 47.
Figure 7. *Cibicidoides semiumbilicatus* (Toutkovski); a,c, opposite sides; b, edge view; × 57.
Figure 8. *Hoeglundina elegans* (d'Orbigny); a,c, opposite sides; b, edge view; × 53.

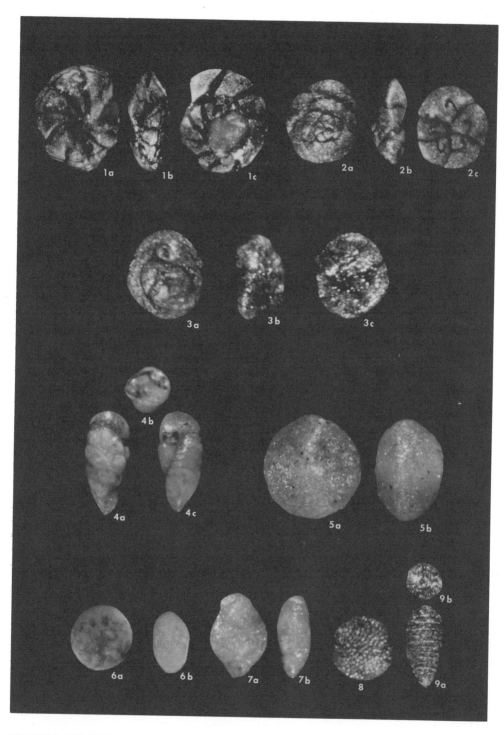

PLATE 6. MAGALLANES BASIN FORAMINIFERA AND RADIOLARIA
Geological Society of America Memoir 139

PLATE 6

Figure 1. *Hoeglundina porcellanea* (Brückmann); a,c, opposite sides; b, edge view; × 48.
Figure 2. *Reinholdella fuenzalidai* Cañon and Ernst, n. sp.; a,c, opposite sides; b, edge view; × 86.
Figure 3. *Reinholdella* cf. *quadrilocula* Subbotina and Datta; a,c, opposite sides; b, edge view; × 84.
Figure 4. *Robertina arctica* d'Orbigny; a,c, opposite sides; b, apertural view; × 61.
Figure 5. Radiolaria, *Spumellaria?* sp. 4; a, side view; b, edge view; × 55.
Figure 6. Radiolaria, *Spumellaria?* sp. 1; a, side view; b, edge view; × 54.
Figure 7. Radiolaria, *Spumellaria?* sp. 5; a, side view; b, edge view; × 76.
Figure 8. Radiolaria, *Spumellaria* sp. 6; side view; × 105.
Figure 9. Radiolaria, *Nassellaria* sp. 3; a, side view; b, top view; × 60.

Selected Bibliography

Andersen, H. V., 1952, *Buccella*, a new genus of the rotalid Foraminifera: Washington Acad. Sci. Jour., v. 42, no. 5, p. 143-151, figs. 1-13.
Andersson, J. G., 1907, Geological fragments from Tierra del Fuego: Uppsala Univ., Geol. Inst. Bull., v. 8, p. 168-183.
Applin, E. R., Ellisor, A. C., and Kniker, H. T., 1925, Subsurface stratigraphy of the coastal plain of Texas and Louisiana: Am. Assoc. Petroleum Geologists Bull., v. 9, p. 79-122, Pl. 3.
Archangelsky, S., 1960a, Estudio anatomico de dos especies del genero *Abietopitys* Krausel, procedentes de la Serie de Nueva Lubecka, Province del Chubut, Argentina: Acta Geol. Lilloana, p. 331-338.
―――1960b, Licopsida y Sphenopsida del Paleozoico Superior del Chubut y Santa Cruz Patagonia: Acta Geol. Lilloana, v. 3, p. 21-36.
Archangelsky, S., and Gamerro, J., 1967, Spore and pollen types of the Lower Cretaceous in Patagonia (Argentina): Rev. Paleobotany and Palynology, v. 1, p. 211-217.
Aubouin, J., 1959, A propos d'un centenaire: Les aventures de la notion de geosynclinal: Rev. Géographie Phys. et Géologie Dynam., v. 2, fasc. 3, p. 135-188.
Auer, Viänö, 1956, The Pleistocene of Fuego-Patagonia, Pt. I: The ice and interglacial ages: Acad. Sci. Fennicae Annales, ser. A. III, Geol.-Geog. 45, 226 p., illus.
―――1958, The Pleistocene of Fuego-Patagonia, Pt. II: History of the flora and vegetation: Acad. Sci. Fennicae Annales, ser. A. III, Geol.-Geog., 239 p., illus.
Barwick, J. S., 1955, The surface stratigraphy of portions of Magallanes province, Chile: Empresa Nacional del Petróleo, Santiago, Chile.
Berry, E. W., 1925, A Miocene flora from Patagonia: Baltimore, Johns Hopkins Univ. Studies in Geology, no. 6, 9 pls.
―――1937, Eogene plants from Rio Turbio in the Territory of Santa Cruz, Patagonia: Baltimore, Johns Hopkins Univ. Studies in Geology, no. 12, p. 11-32, 4 pls.
Bird, J., 1937, Human artifacts in association with horse and sloth bones in southern South America (Magallanes): Science, n.s., v. 86, p. 36-37.
Bonarelli, G., 1917, Informe geológico sobre exploraciones petrolíferas en Magallanes: Santiago, Chile, Ministerio Fomento, Dept. Minas y Petróleo.
Bouma, A. H., 1964, Sampling and treatment of unconsolidated sediments for study of internal structures: Jour. Sed. Petrology, v. 34, no. 2, p. 349.
Brady, H. B., 1884, Report on the scientific results of the voyage of H.M.S. *Challenger:* Zoology, v. 9, p. 814, Pls. 1-115, London, 1884; New York, Johnson Reprint Corp., 1965.
Brandmayr, J., 1945, Contribución al conocimiento geológico del extremo sud-sudoeste del Territorio de Santa Cruz (región Cerro Cazador-alto Rio Turbio): Bol. Inf. Petroleras, v. 12, p. 256.
Brönnimann, P., 1952, Globigerinidae from the Upper Cretaceous (Cenomanian-Maestrichtian) of Trinidad, British West Indies: Bull. Am. Paleontology, v. 34, no. 140, 61 p., 4 pls.
Brückmann, R., 1904, Die Foraminiferen des litauisch-kurischen Jura: Physik.-Ökon. Ges., Königsberg i. Pr., Schr., Jahrg. 45, p. 26, Pl. 4, figs. 17-19.
Brueggen, J., 1929, Communicación preliminar sobre glaciaciones en la Patagonia Austral y Tierra del Fuego: Bol. Minero, v. 45, p. 25-28.

Brueggen, J., 1950, Fundamentos de la geología de Chile: Santiago, Chile, Inst. Geográfico Militar, p. 374.
Caldenius, C. C., 1932, Las glaciaciones cuaternarias en la Patagonia y Tierra del Fuego: Geog. Annaler, v. 14, p. 1-2; Buenos Aires, Dirección Minas y Geología, Pub. no. 95.
Camacho, H. H., 1957, Descripción de una Fauna Marina Paleocena procedente de Tierra del Fuego (Argentina): Ameghiniana, v. 1, nos. 1 and 2, p. 96-100.
Cañon M., A., 1968, Cronestratigrafia de los sedimentos terciarios de Tierra del Fuego: Buenos Aires, Anales Terceras Jornadas Geológicas Argentinas, Tomo I, p. 91-110.
Cecioni, G., 1951, Edad de la Arenisca Springhill: Empresa Nacional del Petróleo, Santiago, Chile.
____1954, Las facies de la Formación Springhill: Empresa Nacional del Petróleo, Santiago, Chile.
____1955a, Edad y facies del Grupo Springhill en Tierra del Fuego: Chile Univ., Fac. Cienc. Fis. y Mat. Anales, v. 12 (Inst. Geol. Pub., no. 6), p. 243-256, 1 fig., 2 pls.
____1955b, Noticias preliminares sobre el hallazgo del Paleozoico Superior en el Archipiélago Patagónico: Chile Univ., Fac. Cienc. Fis. y Mat. Anales, v. 12 (Inst. Geol. Pub., no. 6), p. 257-259, 1 map.
____1955c, Prime notizie sopra l'esistenza del Paleozoico Superiore nell'Arcipelago Patagonico tra i paralleli 50° E. 52° S.: Soc. Toscana Sci. Nat. Mem. (Pisa) Atti, v. 62, ser. A., p. 201-224, 8 figs.
____1955d, Distribuzione verticale di alcun Kossmaticeratidae nella Patagonia cilena: Soc. Geol. Italiana Boll., v. 74, p. 141-149, 1 cuadro.
____1956a, *Leopoldia*(?) *paynensis* Favre: sera posizione stratigrafica in Patagonia: Soc. Italiana Sci. Nat., Milano Atti, v. 95, fasc. 2, p. 135-145, 1 fig.
____1956b, Significato della ornamentazione in alcune Kossmaticeratidae della Patagonia: Rev. Italiana Paleont. Stratig. (Milano), v. 62, no. 1, p. 3-10, tav. 1, Separata, 10 p.
____1957a, Cretaceous flysch and molasse in Departamento Ultima Esperanza, Magallanes province, Chile: Am. Assoc. Petroleum Geologists Bull., v. 41, p. 538-564.
____1957b, Eta della flora del Cerro Guido e stratigrafia, Cile: Soc. Geol. Italiana Boll., v. 76, p. 3-16.
____1959, Sub-Hercynian orogeny in the Strait of Magellan: Soc. Geol. Italiana Boll., v. 78, no. 1.
Cespedes, S., 1963, Reconocimiento geológico del área comprendida entre el Seno Skyring y el Estrecho de Magallanes: Empresa Nacional del Petróleo, Santiago, Chile.
____1964, Reconocimiento geológico en el area Cordillerana al Oeste del Golfo Almirante Montt y hacia el Sur de Isla Dawson: Empresa Nacional del Petróleo, Santiago, Chile.
____1965, Reconocimiento geológico en el Seno Poca Esperanza: Empresa Nacional del Petróleo, Santiago, Chile.
____1968, Estudio geológico en la parte oriental de Isla Hoste: Empresa Nacional del Petróleo, Santiago, Chile.
Charrier, R., 1968, Vinculaciones entre el Terciario inferior marino de Arauco y Magallanes, *in* Cecioni, G., Coordinador, Terciario de Chile—Zona Central: Soc. Geol. Chile, p. 207-209.
Charrier, R., and Lahsen, A., 1965, El límite Cretáceo-Terciario entre el Seno Skyring y el Estrecho de Magallanes, Tesis de Prueba: Santiago, Chile Univ. Fac. Cienc. Fis. y Mat, Dept. Geol.
____1969, Stratigraphy of Late Cretaceous—Early Eocene, Seno Skyring—Strait of Magellan area, Magallanes province, Chile: Am. Assoc. Petroleum Geologists Bull., v. 53, no. 3, p. 568-590, 8 figs., 3 tables.
Cole, W. S., 1938, Stratigraphy and micropaleontology of two deep wells in Florida: Florida Geol. Survey Bull., no. 16, p. 45, Pl. 3, fig. 5.
Collignon, M., 1967, *Ammonites* du Chili (Magellanie—Terre de Feu) Communiquées par J. Sigal: Inst. Français Pétrole.
Cookson, I. C., and Cranwell, L. M., 1967, Lower Tertiary microplankton. Spores and pollen grains from southernmost Chile: Micropaleontology, v. 13, no. 2, p. 204-216.
Cortes, R., 1960, Revisión de algunos problemas geológicos en parte de Ultima Esperanza y las expectativas petrolíferas de la región: Empresa Nacional del Petróleo, Santiago, Chile.

___1964, Estratigrafía y un estudio de paleocorrientes del flysch Cretaceo del Depto. de Ult. Esperanza, Tesis de grado presentada: Santiago, Univ. Tec. del Estado.

___1968, Reconocimiento geológico preliminar en la región de Isla Navarino y Bahía Tekenika: Empresa Nacional del Petróleo, Santiago, Chile.

___1969, Reconocimiento geológico de algunos afloramientos del Paleozoico, Jurasico y Cretaceo Inferior en la parte N de la Provincia de Santa Cruz (Argentina): Empresa Nacional del Petróleo, Santiago, Chile.

Cortes, R., and Cañon, A., 1964, Rango estratigráfico y frecuencia de los principales fósiles del Cretaceo de Ultima Esperanza: Empresa Nacional del Petróleo, Santiago, Chile.

Cushman, J. A., 1926, The Foraminifera of the Velasco shale of the Tampico embayment: Am. Assoc. Petroleum Geologists Bull., v. 10, p. 581-612, Pls. 15-21.

___1927, American Upper Cretaceous species of *Bolivina* and related species: Cushman Found. Foram. Research Contr., v. 2, p. 85-91, Pl. 12.

___1930, The Foraminifera of the Atlantic Ocean, Pt. 7: U.S. Natl. Mus. Bull. 104, 79 p., 18 pls.

___1938a, Cretaceous species of *Gümbelina* and related genera: Cushman Found. Foram. Research Contr., v. 14, pt. 1, p. 1-30, Pls. 1-4.

___1938b, Additional new species of American Cretaceous Foraminifera: Cushman Found. Foram. Research Contr., v. 14, pt. 2, p. 31-50, Pls. 5-8.

___1939, A monograph of the foraminiferal family Nonionidae: U.S. Geol. Survey Prof. Paper 191, 100 p., 20 pls.

___1950, Foraminifera, their classification and economic use: Cambridge, Mass., Harvard Univ. Press, 4th ed., 605 p., 55 pls.

___1951, Paleocene Foraminifera of the Gulf coastal region of the United States and adjacent areas: U.S. Geol. Survey Prof. Paper 232, 75 p.

Cushman, J. A., and Bermudez, P. J., 1936, Additional new species of Foraminifera and a new genus from the Eocene of Cuba: Cushman Found. Foram. Research Contr., v. 12, p. 55-63, Pls. 10, 11.

Cushman, J. A., and Parker, F. L., 1947, *Bulimina* and related foraminiferal genera: U.S. Geol. Survey Prof. Paper 210-D, p. 55-176, Pls. 15-30.

Cushman, J. A., and Renz, H. H., 1942, Eocene, Midway, Foraminifera from Soldado Rock, Trinidad: Cushman Found. Foram. Research Contr., v. 18, pt. 1, p. 1-22, Pls. 3, 4.

___1946, The foraminiferal fauna of the Lizard Springs formation of Trinidad, British West Indies: Cushman Found. Foram. Research Spec. Pub. no. 18, 48 p, 8 pls.

Cushman, J. A., and Todd, R., 1943, The genus *Pullenia* and its species: Cushman Found. Foram. Research Contr., v. 19, pt. 1, p. 1-23, Pls. 1-4.

___1949, Species of the genera *Allomorphina* and *Quadrimorphina:* Cushman Found. Foram. Research Contr., v. 25, pt. 3, p. 59-72, Pls. 11, 12.

Darwin, Ch. R., 1846, Geological observations on South America: London, Smith and Elder, 279 p., 5 pls.

Decat, J., and Pomeyrol, 1931, Informe geológico sobre las posibilidades petrolíferas de la región de Magallanes: Bol. Minas y Petróleo, v. 1, no. 9.

Di Persia, C. A., 1960, A cerca del descubrimiento del Precámbrico en la Patagonia extra-andina (Prov. de Chubut): Anales Primeras Jornadas Geologicas Argentinas, v. 2, p. 65-68.

Duhart, J., 1963, Estudio geológico del área Seno Almirantazgo Canal Whiteside-Lago Blanco-Seno Almirantazgo, Tierra del Fuego, Magallanes, Tesis de Prueba: Santiago, Chile Univ. Fac. Cienc. Fis. y Mat.

Edgell, H. S., 1957, The genus *Globotruncana* in northwest Australia: Micropaleontology, v. 3, p. 101-122, Pls. 1-4.

Espitalie, J., and Sigal, J., 1963, Contribution à l'étude des Foraminifères (micropaléontologie-microstratigraphie) du Jurrassique supérieur et Majunga (Madagascar): Annales Géol., Madagascar, fasc. 32, p. 33, Pl. 10, figs. 6, 7.

Favre, F., 1908, Description d'une faune d'Ammonites du Crétacique inférieur de Patagonie: Archives Sci., v. 27, p. 167-190.

Felsch, J., 1913, Informe sobre el reconocimiento geológico de los alrededores de Punta Arenas y de la parte noroeste de Tierra del Fuego con el objeto de encontrar posibles yacimientos de petróleo: Santiago, Chile, Bol. Soc. Nac. Mineria.

___1916, Reconocimiento geológico de los terrenos petrolíferos de Magallanes del Sur: Santiago, Chile, Bol. Soc. Nac. Mineria, p. 214-223, p. 309-315.

Feruglio, E., 1936-1937, Paleontographia Patagónica: Padova Mem. Inst. Geol., v. 11-12, p. 384.
____1939, Mapa geológico de la Patagonia al sur del paralelo 42° y Tierra del Fuego: Bol. Inf. Petroleras.
____1942, Recientes progresos en el conocimiento geológico de la Patagonia y Tierra del Fuego: Anales Primer Congreso Panamer, Ingenieria Minas y Geología, v. 2, p. 380-401.
____1949-1950, Descripcion geológica de la Patagonia: Bol. Inf. Petroleras, Tomo 1, 2, 3.
____1951, Su alcune piante del Gondwana inferiore della Patagonia: Torino Univ. Inst. Geol. Pub., v. 1, p. 1-34.
____1952, Estado actual del conocimiento geológico de la Patagonia, Republic Argentina: Internat. Geol. Cong., 18th, London 1948, pt. 5, p. 278.
Finlay, H. J., 1939, New Zealand Foraminifera: Key species in stratigraphy—No. 2: Royal Soc. New Zealand Trans. Proc., v. 69, pt. 1, p. 89-128, Pls. 11-14.
____1940, New Zealand Foraminifera: Key species in stratigraphy—No. 4: Royal Soc. New Zealand Trans. Proc., v. 69, pt. 4, p. 448-472, Pls. 62-67.
Frizzell, D. L., 1943, Upper Cretaceous Foraminifera from northwestern Peru: Jour. Paleontology, v. 17, p. 331-353, Pls. 55-57.
____1954, Handbook of Cretaceous Foraminifera of Texas: Texas Univ. Bur. Econ. Geol. Rept. Inv. no. 22, p. 232.
Fuenzalida, H., 1942, El Magallanico de la Isla Riesco con referencias a algunas regiones adyacentes: Primer Congreso Panamer, Ingeniería Minas y Geol., p. 402-428.
____1964, Las faunas del geosinclinal Andino y del geosinclinal de Magallanes: Soc. Geol. Chile, no. 2, p. 1-27.
Garcia, E. Rossi De, and Camacho, H. H., 1965, Descripción de fósiles procedentes de una perforación efectuada en la provincia de Santa Cruz (Argentina): Ameghiniana, v. 4, no. 3, p. 71-73.
Gonzalez, E., 1953, Estratigrafía y distribución de los grupos El Salto y Palomares en gran parte de la Cuenca de Magallanes: Empresa Nacional del Petróleo, Santiago, Chile.
____1954, Petrologia de la Arenisca Springhill en la parte oriental de la Cuenca de Magallanes: Empresa Nacional del Petróleo, Santiago, Chile.
____1965, La Cuenca petrolífera de Magallanes: Apartado Rev. Minerales, no. 91, p. 1-15.
Groeber, P., 1952, Geografía de la República Argentina: Soc. Argentina Estudios Geografícos, v. 2, pt. 1, p. 1-541.
Grossling, B., 1953a, El petróleo de Magallanes: Empresa Nacional del Petróleo, Santiago, Chile.
____1953b, Geologia de petróleo de la Formación Springhill en el Distrito Springhill, Magallanes: Anales del Inst. Ingeniería de Chile, 1966, no. 7-8, p. 184-197; no. 9-10, p. 225-242; no. 11-12, p. 247-269.
Gutierrez, A., 1962, Estudio de las rocas basales de la Cuenca de Magallanes, Tesis de Prueba: Santiago, Chile Univ. Fac. Cienc. Fis. y Mat.
Hada, Y., 1931, Notes on the Recent Foraminifera from Mutsu Bay, *in* Report of the biological survey of Mutsu Bay, no. 19: Tohoku Imp. Univ. Sci. Repts., ser. 4 (biol.), v. 6, p. 120, Pl. 121, figs. 79a-c.
Halle, Th. G., 1912, On the occurrence of Dictyozamites in South America (Tierra del Fuego): Paläobot. Zeitschr., v. 1, p. 40-42.
____1913, Some Mesozoic plant-bearing deposits in Patagonia and Tierra del Fuego and their floras: Stockholm, Kungl. Svenska Vet. Akad. Handl. 60, p. 3, 5 pls.
Halpern, M., 1962, Potassium-argon dating of plutonic bodies in Palmer Peninsula and southern Chile: Science, v. 138, p. 1261-1262.
____1967, Geologic significance of isotopic age measurements of rocks from Tierra del Fuego, Chile: Dallas, Texas, Southwest Center for Advanced Studies.
Harrington, H. J., 1943, Observaciones geológicas en la Isla de los Estados: Mus. Argentino Cienc. Nat. Anales, v. 41, p. 29-52.
____1962, Paleogeographic development of South America: Am. Assoc. Petroleum Geologists Bull., v. 46, p. 1773-1814.
____1965, *in* Ludwig, W., Ewing, J., and Ewing, M., Seismic refraction measurements in the Magallanes Straits: Jour. Geophys. Research, v. 70, no. 8, p. 1855-1876.
Hatcher, J. B., 1903, Reports of the Princeton expeditions to Patagonia, 1896-1899. I. Narrative of the expeditions, geography of southern Patagonia: Princeton, N.J., Princeton Univ. Press.

Hauser, A., 1964, La "Zona Glauconitica" en la Plataforma Springhill, Magallanes, Tesis de Prueba: Santiago, Chile Univ. Escuela Geología.

Hauthal, R., 1899, Sur le crétace et le tertiaire de la Patagonie australe: Buenos Aires, Rev. Mus. La Plata, v. 10, p. 43-45.

Hauthal, R., Wilckens, O., and Paulcke, W., 1907, Die obere Kreide Südpatagonicus und ihre Fauna: Ber. Naturf. Gesell., Freiburg, v. 15, p. 75-248, Pls. 1-19.

Hedberg, H. D., 1937, Foraminifera of the middle Tertiary Carapita formation of northeastern Venezuela: Jour. Paleontology, v. 11, p. 661-697, Pls. 90-92.

Hemmer, A., 1936, Resultados obtenidos de las exploraciones geológicas de la region de Magallanes desde Noviembre de 1932 hasta Enero de 1934: Bol. Minas y Petróleo, v. 5, no. 36.

_____1937, Las exploraciones petroliferas en Magallanes: Bol. Minas y Petróleo, v. 5, no. 77.

Herm, D., 1966, Micropaleontological aspects of the Magellanese geosyncline, southernmost Chile, South America: Proc. 2d, West African Micropaleontological Collection (Ibadan, 1965), p. 72-86.

Hoffstetter, R., Fuenzalida, H., and Cecioni, G., 1957, Lexique stratigraphique international-Amérique Latine, fasc. 7, Chile: Centre National Recherche Scientifique, Paris, 7, p. 1-444.

Hollingsworth, R. V., 1954, Preliminary report on Fusulinids from Patagonia-Paleontological Laboratory: R. V. Hollingsworth, Midland, Texas.

Hornibrook, N., 1958, New Zealand Upper Cretaceous and Tertiary foraminiferal zones and some overseas correlations: Micropaleontology, v. 4, no. 1, p. 25-38.

Hyades, P.D.J., 1887, Geologie, in Mission scientifique du Cap Horn, 1882-1883: Paris, Gauthier-Villars, v. 4, p. 1-242.

Katz, H. R., 1961a, Algunas notas acerca de la intrusión granítica en la Cordillera Paine-Provincia de Magallanes: Rev. Minerales, v. 74, p. 1-15.

_____1961b, Sobre la ocurrencia de Cretaceo Superior Marino en Coyhaique, Provincia de Aisen: Santiago, Chile Univ. Fac. Cienc. Fis. y Mat. Anales, Pub. no. 21, p. 113-128.

_____1961c, Descubrimiento de una microflora Neocomiana en la Formación Agua Fresca (Eoceno) de Magallanes y su significado con respecto a la evolución tectónica de la zona: Santiago, Chile Univ. Fac. Cienc. Fis. y Mat. Anales, Pub. no. 21, p. 133-141.

_____1962, Fracture patterns and structural history in the sub-Andean belt of southernmost Chile: Jour. Geology, v. 70, p. 595-603.

_____1963, Revisión of Cretaceous stratigraphy in Patagonian Cordillera of Ultima Esperanza, Magallanes province, Chile: Am. Assoc. Petroleum Geologists Bull., v. 47, p. 506-524.

_____1964, Conceptos nuevos sobre el desarrollo geosinclinal y del sistema cordillerano en el extremo austral del continente. Resumenes: Soc. Geol. Chile, v. 7, p. 1-8.

Katz, H. R., and Watters, W. A., 1965, Geological investigations of the Yahgan Formation (upper Mesozoic) and associated igneous rocks of Navarino Island, southern Chile, New Zealand: Jour. Geology and Geophysics, v. 9, no. 3, p. 323-359.

Keidel, J., and Hemmer, A., 1931, Informe preliminar sobre las investigaciones efectuadas en la region petrolífera de Magallanes en los meses de verano de 1928-1929: Bol. Minero, v. 47, p. 706-717.

Kniker, H. T., 1947, Age determinations of Tertiary formations of Peninsula Brunswick by means of Foraminifera: Empresa Nacional del Petróleo, Santiago, Chile.

_____1949, Present information regarding ages of the Tertiary formations of Magallanes province: Empresa Nacional del Petróleo, Santiago, Chile.

_____1950a, Some derivations of specific names for Foraminifera: Empresa Nacional del Petróleo, Santiago, Chile.

_____1950b, Notes on formations, determinations of sediments of Peninsula Brunswick and Tierra del Fuego: Empresa Nacional del Petróleo, Santiago, Chile.

Kranck, E. H., 1932, Geological investigations in the Cordillera of Tierra del Fuego: Acta Geog., v. 4, no. 2.

_____1933, Sur quelques roches à Radiolaires de la Terre de Feu: Soc. Géol. France Bull., ser. 5, v. 2, p. 275-283, Pl. 16.

Leanza, A. F., 1963, Patagoniceras gen. nom. *Binneyitidae* y otros amonites del Cretaceo Superior de Chile Meridional con notas acerca de su posición estratigrafíca: Cordova, Argentina, Separata Bol. Academia de Ciencias, Tomo 43, ent. 2a, 3a, 4a.

Levi, B., Meregh, S., and Munizaga, F., 1963, Edades radiométricas y petrografía de granitos Chilenos: Chile Inst. Inv. Geol. Bol. no. 12.

Loeblich, A. F., and Kniker, H. T., 1950, The foraminiferal genus *Rzehakina* in Magallanes province, Chile: Empresa Nacional del Petróleo, Santiago, Chile.

Loeblich, A. R., Jr. and Tappan, H., 1964, Foraminiferida, *in* Moore, R. C., ed., Treatise on invertebrate paleontology, Pt. C, Protista 2: Lawrence, Kansas, Univ. Kansas Press and Geol. Soc. America, p. C55-C900, figs. 34-642.

Ludwig, W. J., 1965, Seismic refraction measurement in the Magellan Straits: Jour. Geophys. Research, v. 70, p. 1855-1876.

Mallory, V. S., 1959, Lower Tertiary biostratigraphy of the California Coast Ranges: Am. Assoc. Petroleum Geologists Pub. 614, p. 297, 42 pls., 18 tables.

Malumian, N., 1969, Foraminíferos del Cretácico Superior y Terciario del subsuelo de la Provincia Santa Cruz, Argentina: Ameghiniana, v. 5, no. 6, p. 191-227.

Marks, P., Jr., 1951. A revision of the smaller Foraminifera from the Miocene of the Vienna basin: Cushman Found. Foram. Research Contr., v. 2, pt. 2, p. 33-73, Pls. 5-8.

Martinez, R., 1954a, Revision de las especies de Elphidium de la Cuenca Magallanica: Empresa Nacional del Petróleo, Santiago, Chile.

―――1954b, Revisión de algunas Lagenidae de la Cuenca Magallanica: Empresa Nacional del Petróleo, Santiago, Chile.

―――1957, El Terciario Superior de algunos pozos del Continente y Tierra del Fuego: Empresa Nacional del Petróleo, Santiago, Chile.

―――1958, Consideraciones morfológicas para el estudio de los foramíniferos: Empresa Nacional del Petróleo, Santiago, Chile.

―――1964, *Bolivinoides draceo dorreeni* Finlay from the Magellan Basin, Chile: Micropaleontology, v. 2, p. 360-364.

―――1968, Zonación preliminar del Terciario de Chile central mediante foraminíferos planctónicos y su correlación regional y transcontinental, *in* El Terciario de Chile, zona central: Soc. Geol. Chile, p. 191-203.

Martínez, R., and Ernst, M., 1960, Informe preliminar sobre la edad las formaciones Cretacicas deala Cuenca Magallanica: Empresa Nacional del Petróleo, Santiago, Chile.

Martínez, R., Osorio, R., and Lillo, J., 1964, Estudio micropaleontológico del miembro Ciervos de la Formación Loreto, Magallanes: Santiago, Chile Univ. Fac. Cienc. Fis. y Mat.

Menendez, C. A., 1965, Microplancton fósil de sedimentos terciarios y cretáceos del norte de Tierra del Fuego (Argentina): Ameghiniana, v. 4, no. 1, p. 7-15.

Mercerat, A., 1891, Sinopsis de la familia de los Astrapotheridae: Buenos Aires, Rev. Mus. La Plata, v. 1, p. 237-256.

Montadert, M., 1968, *in* Institut Français de Pétrole, Etude des possibilités pétrolières de la plateforme Springhill (Bassin de Magellan): Rapport principal, Empresa Nacional del Petróleo, Santiago, Chile, p. 44-46.

Mordojovich, C., 1951, The micropaleontological laboratory in Punta Arenas, Chile: Micropaleontology, v. 5, no. 4, p. 10-12.

Morrow, A. L., 1934, Foraminifera and Ostracoda from the Upper Cretaceous of Kansas: Jour. Paleontology; v. 8, p. 186-205, Pls. 29-31.

Munoz Cristi, J., 1950, *in* Geografía económica de Chile: Santiago Corporación Fomento Producción, no. 1, p. 5-187.

―――1956, Chile, *in* Jenks, W. F., ed., Handbook of South American geology: Geol. Soc. America Mem. 65, p. 189-214.

Myatliuk, E. V., 1939, Foraminifera from the Upper Jurassic and Lower Cretaceous deposits of the middle Volga region and Obshchyi Syrt. Neftianyi: Leningrad, Geologo-razvedochnyi Inst., Trudy (Trans. Oil Geol. Inst.), ser. A, fasc. 120, p. 56 (Russian) or p. 72 (English), Pl. 4, figs. 41-42.

Natland, M. L., 1933, Temperature and depth ranges of some Recent and fossil Foraminifera in the southern California region: Scripps Inst. Oceanog. Bull., Tech. Ser., v. 3, no. 10, p. 225-230, 1 table.

―――1950, Report on the Pleistocene and Pliocene Foraminifera: Geol. Soc. America Mem. 43, p. 1-55.

―――1957, Paleoecology of west coast Tertiary sediments, *in* Ladd, H. S., ed., Treatise on marine ecology and paleoecology, Vol. 2: Geol. Soc. America Mem. 67, p. 543-571.

―――1967, New classification of water-laid clastic sediments: Am. Assoc. Petroleum Geologists

Bull., v. 53, no. 3, p. 476.
Natland, M. L., and Kuenen, P. H., 1951, Sedimentary history of the Ventura basin and the action of turbidity currents: Soc. Econ. Paleontologists and Mineralogists Spec. Pub. no. 2, p. 76-107.
Natland, M. L., and Gonzalez, E., 1965, A system of stages for correlation of Magallanes Basin sediments: Empresa Nacional del Petróleo, Santiago, Chile, p. 1-18.
Nordenskjold, O., 1897, Notes on Tierra del Fuego. An account of the Swedish expedition of 1895-1897: Edinburgh Geog. Mag., v. 13, p. 393-399.
Nuttall, W.L.F., 1930, Eocene Foraminifera from Mexico: Jour. Paleontology, v. 4, p. 271-293, Pls. 23-25.
Ortmann, A., 1898, Preliminary report on some new marine Tertiary horizons discovered by J. B. Hatcher near Punta Arenas: Am. Jour. Sci., v. 6, p. 478-482.
_____1899, The fauna of the Magellanian beds of Punta Arenas, Chile: Am. Jour. Sci., v. 8, p. 427-432.
_____1900, Synopsis of the collections of invertebrate fossils made by the Princeton expedition to southern Patagonia: Am. Jour. Sci., v. 10, p. 368-381.
_____1902, Report of the Princeton University expedition to Patagonia, Tertiary invertebrates: Princeton, N.J., Princeton Univ. Press, v. 4, no. 2, p. 45-332, 28 pls.
Owen, R., 1840, The zoology of the voyage of H.M.S. *Beagle*, Pt. 1, Fossil Mammalia: London, Smith, Elder and Co., p. 111, 32 pls.
Paulcke, W., 1907, Die Cephalopoden der oberen Kreide Südpatagoniens: Ber. Naturf. Gesell., Freiburg, v. 15, p. 167-248, 10 pls.
Perebaskine, V., 1935, Note préliminaire sur les foraminifères provenant d'un sondage foré dans la région du Détroit de Magellan (Chili): Soc. Géol. France, Comptes Rendus, fasc. 1, p. 104-105.
Petters, V., and Sarmiento, R., 1956, Oligocene and lower Miocene biostratigraphy of the Carmen-Zambrano area, Colombia: Micropaleontology, v. 2, no. 1, p. 7-35.
Phillippi, R. A., 1896, Patagonia and Chile; their orography and geology contrasted: Scottish Geog. Mag., v. 12, p. 303.
Plummer, H. J., 1926, Foraminifera of the Midway Formation in Texas: Texas Univ. Bull., no. 2644, 206 p., 15 pls., 13 figs., chart.
Poag, C. W., 1966, Paynes Hammock (lower Miocene?) Foraminifera of Alabama and Mississippi: Micropaleontology, v. 12, p. 393-440, Pls. 1-9.
Quensel, P. D., 1911, Geologisch-petrographische Studien in der patagonischen Cordillera: Uppsala Univ. Geol. Inst. Bull., v. 11, p. 1-114.
_____1913, Die Quarzporphyr und Porphyroidformation in Südpatagonien und Feuerland: Uppsala Univ. Geol. Inst. Bull., v. 12, p. 9-40.
Reeside, J. B., Jr., 1950, Report on referred fossils (Upper Cretaceous and Tertiary invertebrates, Magallanes province): Empresa Nacional del Petróleo, Santiago, Chile.
Reuss, A. E., 1863, Die Foraminiferen des norddeutschen Hils und Gault: K. Akad. Wiss., Wien, Math.—Naturw. cl., Sitzber., v. 46 (1862), pt. 1, p. 5-100, Pls. 1-13.
Richter, M., 1925, Beiträge zur Kenntnis der Kreide in Fuerland: Neues Jahrb Mineralogie Geologie u. Paläontologie, v. 52, p. 529-568, Pls. 6-9.
Robles, M. L., Gomez, P. M., and Arellano, A.R.V., 1956, Foraminiferos del Cretacico Superior y Paleoceno de la Provincia de Magallanes, Chile: Internat. Geol. Cong., 20th, Mexico [D.F.] 1937, p. 184-185.
Ruby, Glen M., 1944, The search for oil in Chile: Petróleo Interamericano, Septiembre-Octubre.
Scott, K. M., 1964, Sedimentology of conglomeratic mud flows in a flysch sequence, Chilean Patagonia: Geol. Soc. America, Abs. for 1963, Spec. Paper 76, p. 146.
_____1966, Sedimentology and dispersal pattern of a Cretaceous flysch sequence, Patagonian Andes, southern Chile: Am. Assoc. Petroleum Geologists Bull., v. 50, no. 1, p. 72-107.
Scott, W. B., 1903, Reports of the Princeton University expeditions to Patagonia, 1896-1899. Vol. 5, Paleontology. Mammalia of the Santa Cruz beds, Pt. I. Edentata: I. Dasypoda, p. 1-106; II. Glyptodontia and Gravigrada, p. 107-227; III. Gravigrada, p. 227-364. Pt. II. Insectivors: p. 365-383; Pt. III. Glires, p. 365-499: Princeton, N.J., Princeton Univ. Press.
_____1905, The Mammalian fauna of the Santa Cruz beds: Internat. Zool. Cong., 6th, Bern 1904, Comptes Rendus, p. 241-266.
_____1928, Reports of Princeton University expeditions to Patagonia 1896-1899, Astrapotheria

of the Santa Cruz beds: Princeton, N.J., Princeton Univ. Press. v. 6, p. 301-342.
Severin, E., 1951a, Formacion Brush Lake, ubicación gráficos de distribución y láminas: Empresa Nacional del Petróleo, Santiago, Chile.
──1951b, Formación Agua Fresca, gráfico de correlaciones y láminas: Empresa Nacional del Petróleo, Santiago, Chile.
──1951c, Formación Puerto Nuevo plano de ubicación y muestreo, cuadros de distribución y frecuencia. Láminas: Empresa Nacional del Petróleo, Santiago, Chile.
──1955, Correlaciones micropaleontológicas en el Cretáceo y en el Terciario: Empresa Nacional del Petróleo, Santiago, Chile.
Sigal, J., 1952, Aperçu stratigraphique sur la micropaléontologie du Crétace: Internat. Geol. Cong., 19th, Algérie, Mon. Region., ser. k., no. 26, p. 1-47, figs. 1-46, table.
──1967, Estudio coronestratigráfico sobre la Formación Springhill y los Estratos con Favrella en los pozos de Magallanes y Tierra del Fuego: Empresa Nacional del Petróleo, Santiago, Chile.
Sigal, Jacques, Grekoff, Nicolas, Singh, N. P., Cañon, A., and Ernst, M., 1970, Sur l'âge et les affinités "gondwaniennes" de microfaunes (foraminifères et ostracodes) malgaches, indiennes, et chiliennes au sommet du jurassique et a la du crétace: Acad. Sci. Comptes Rendus, v. 271, ser. D, no. 1, p. 24-27.
Simpson, G. G., 1940, Review of the mammal-bearing Tertiary of South America: Am. Philos. Soc. Proc., v. 83, no. 5, p. 649-709.
──1941, A Miocene sloth from southern Chile: Am. Philos. Soc. Proc., no. 1156.
Sinclair, W. J., 1901-1906, Reports of Princeton University expedition to Patagonia, Mammalia of the Santa Cruz beds, Vol. IV, Pt. III (Marsupialia). Vol. VI, Pt. I (typotheria): Princeton, N.J., Princeton Univ. Press.
Sociedad Geologica de Chile, 1968, El Terciario de Chile—Zona Central, Cecioni, G., Coordinador: Santiago, p. 1-280, 5 pls.
Steffen, H., 1909-1910, Viajes de exploración y estudio en la Patagonia occidental 1892-1902, 2 tomos: Santiago, Chile, Imprenta Cervantes.
Steinmann, G., 1908, Das Alter der Schieferformation im Feuerlande: Mineralogie, Geologie, und Paleontologie, p. 193-194.
Stipanicic, P., and Reig, O., 1955, Breve noticia sobre el hallazgo de anuros en el denominado Complejo Porfírico de la Patagonia extra-Andina, con consideraciones acerca de la composicion geológica del mismo: Asoc. Geol. Argentina Rev., v. 10, no. 4.
Stolley, E., 1912, Über einige Cephalopoden aus der unteren Kreide Patagoniens: Uppsala, Stockholm, Archiv. Zool., v. 7, p. 23.
Subbotina, N. N., Datta, A. K., and Srivastava, B. N., 1960, Foraminifera from the Upper Jurassic deposits of Rajasthan (Jaisalmer) and Kutch, India: India Geol. Min. Metal. Soc. Bull., no. 23, p. 39, Pl. 3, figs. 8a-c.
Suero, T., 1948, Descubrimiento de Paleozoico superior en la zona extrandina de Chubut. Nota preliminar: Bol. Inf. Petroleras, v. 25, no. 287.
──1953, Las sucesiones sedimentarias suprapaleozoicas de la zona extrandina del Chubut: Asoc. Geol. Argentina Rev., v. 8, no. 1.
──1958, Datos geológicos sobre el Paleozoico superior en la zona de Nueva Lubecka y alrededores (Chubut), (Extrandino) Provincia del Chubut: Rev. Museo de La Plata (Nueva serie), Geología, no. 30.
──1962, Paleogeografía del Paleozoico superior en la Patagonia (Republic Argentina): Asoc. Geol. Argentina Rev., v. 16, no. 1, p. 2.
Suero, T., and Criado, R., 1955, Descubrimiento del Paleozoico Superior al oeste de la Bahía Laura (Terr. Nac. de Santa Cruz) y su importancia paleogeográfica: Notas del Museo de La Plata, v. 18, Geología, no. 68, p. 157-168.
Tappan, H., 1940, Foraminifera from the Grayson Formation of northern Texas: Jour. Paleontology, v. 14, p. 93-126, Pls. 14-19.
──1955, Foraminifera from the Arctic Slope of Alaska, Pt. 2, Jurassic Foraminifera: U.S. Geol. Survey Prof. Paper 236-B, p. 21-90, Pls. 7-28.
ten Dam, A., and Sigal, J., 1950, Some new species of Foraminifera from the Dano-Montian of Algeria: Cushman Found. Foram. Research Contr., v. 1, pts. 1, 2, p. 31-37, Pl. 2.
Terquem, O., 1866, Sixième mémoire sur les foraminifères du Lias des départementes de l'Indre et de la Moselle: Metz, Lorette, p. 459-532, Pls. 19-22.

Thomas, C. R., 1949a, Manantiales field, Magallanes province, Chile: Am. Assoc. Petroleum Geologists Bull., v. 33, no. 9.
_____1949b, Geology and petroleum exploration in Magallanes province, Chile: Am Assoc. Petroleum Geologists Bull., v. 33, no. 9, p. 1553-1568.
Todd, Ruth, and Kniker, H. T., 1952, An Eocene foraminiferal fauna from the Agua Fresca shale of Magallanes province, southernmost Chile: Cushman Found. Foram. Research Spec. Pub. no. 1, Washington, 28 p., 4 pls.
Toenges, A. L., 1948, Coals of Chile: U.S. Bur. Mines Bull. 474.
Toutkovski, P., 1887, Foraminifera of the Tertiary and Cretaceous deposits of Kiev (Russian): Soc. Nat. Kieff Mem., Russie, v. 8, no. 2, p. 358, Pl. 5, fig. 5.
Trujillo, E. F., 1960, Upper Cretaceous Foraminifera from near Redding, Shasta County, California: Jour. Paleontology, v. 34, p. 290-346, Pls. 43-50.
Tyrrell, G. W., 1932, The basalts of Patagonia: Jour. Geology, v. 40, p. 374-383.
Ugarte, F., 1966, La cuenca compuesta Carbonífero-Jurásica de la Patagonia Meridional: Univ. Patagonia San Juan Bosco Anales, Ciencias geológicas, Tomo I, no. 1, p. 37-68, 8 figs.
Vieaux, D. G., 1941, New Foraminifera from the Denton Formation in northern Texas: Jour. Paleontology, v. 15, p. 624-628, Pl. 85.
Watters, W. A., 1965, Prehnitization in the Yahgan Formation of Navarino Island, southernmost Chile (C. E. Tilley complimentary volume): Mineralog. Mag., p. 517-527.
Weaver, Ch. E., 1931, Paleontology of the Jurassic and Cretaceous of west central Argentina: Washington Univ. Mem., I, p. 1-469, 62 pls.
Weeks, L. G., 1947, Paleogeography of South America: Am. Assoc. Petroleum Geologists Bull., v. 31, p. 1194.
Weiss, L., 1955, Planktonic index Foraminifera of northwestern Peru: Micropaleontology, v. 1, no. 4, p. 301-319.
Wenzel, O., 1951, Conocimiento actual sobre la geología de la provincia de Magallanes y sus posibilidades petroleras: An. Inst. Ing. Chile, v. 44, p. 202-213.
White, C., 1890, On certain Mesozoic fossils from the islands of St. Paul and St. Peter in the Strait of Magellan: U.S. Natl. Mus. Proc., v. 13, p. 13-14.
Wicher, C. A., 1949, On the age of the higher Upper Cretaceous of the Tampico Embayment area of Mexico: Belgrade, Mus. Hist. Nat. Pays Serbe Bull., ser. A, no. 2, p. 57 (Serbian); p. 85 (English), Pl. 5, figs. 2, 3.
Wilckens, O., 1904, Über Fossilien der oberen Kreide Südpatagoniens, Centralblatt f.: Mineralogie, Geologie und Paläeontologie, p. 597-599.
_____1905, Die Meeresablagerungen der Kreide- und Tertiarformation in Patagonien: Mineralogie, Geologie und Paläeontologie, v. 21, p. 98-195.
_____1907, Erläuterung zu R. Hauthals geologischer Skizze des Gebietes zwischen dem Lago Argentino und dem Seno de la Ultima Esperanza (Südpatagonien): Naturf. Gesell. Freiburg, v. 15, p. 75-96, 1 pl.
_____1921, Beiträge zur Paläontologie von Patagonien. Mit einem Beitrag von G. Steinmann: Neues Jahrb. Mineralogie, Geologie und Paläontologie.
_____1924, Zur Stratigraphie von Patagonien: Geol. Rundschau, v. 15, p. 315-317.
Windhausen, A., 1929-1931, Geologia Argentina, 2 Vols.: Buenos Aires, J. Peuser.
Woodward, A. Smith, 1900, On some remains of Grypotherium (Neomylodon) Listai and associated mammals from a cavern near Consuelo cove (Felis, Archtotherium): Zool. Soc. Proc., p. 64-79.
Zeil, W., 1958, Sedimentation in der Magallanes-Geosinklinale mit besonderer Berücksichtigung des Flysch: Geol. Rundschau, v. 47, p. 425-443.

MANUSCRIPT RECEIVED BY THE SOCIETY AUGUST 10, 1971
REVISED MANUSCRIPT RECEIVED MAY 11, 1973

Index

Abuillot formation, 72
Africa, 13, 14
Agua Fresca formation, 63
Agua Fresca River, 35, 37, 79
Albian, 3, 38, 43
Algal deposits, 43
Allomorphina conica, 37, 85, 100, 101
Almirante Martinez area, 48
Ammobaculites barrowensis, 46, 68, 94, 95
Andean foothill belt, 9
Andean intrusive suite, 9, 15
Andean Island Arc system, 15
Andean suite, 9
Anomalina popenoei, 84
Anomalina rubignosa, 37, 88, 102, 103
Antartica-India-Australia continental mass, 13
Anura sp., 48
Aptian, 3, 43
Apticus sp., 46
Archangelsky, 10
Archipelago area, 10
Archipelago Mountain Range, 9
Argentina, 10, 11, 12, 18
Astacolus
  cf. *crepidula*, 73
  filosa, 46
  microdictyotos, 13, 46, 73, 94, 95
  skyringensis, 73, 96, 97
  stillus, 46, 73, 96, 97
  tricarinellus, 46, 74, 96, 97
Aulacosphinctes, 46

Bahía Fuentes (Brunswick Peninsula), 67, 88
Bahía Inútil, 28, 31, 73, 76, 77, 85, 87
Bahía L'aura, 51-57
Ballena Formation, 16, 56
Ballena Hills, 29, 33
Bandurria, 25
Barremian, 3, 13, 43

Bartonian, 82
Beach deposits, 16
Belemnopsis patagoniensis, 46
Bellota Creek, 48, 49
Bermudez, 29
Bertrand Island, 14
Blanco River, 37, 39
Bolivina incrassata, 37, 76, 96, 97
  v. *crassa*, 76
  v. *gigantea*, 38, 76, 96, 97
  v. *lata*, 76
  reussi v. *navarroensis*, 76
Bolivinoides draceo v. *dorreeni*, 38, 64
Brotzen, 72
Brunswickian Stage, 3, 15, 29, 34, 56, 69, 70, 72, 73, 75, 79, 82, 83
Brunswick Peninsula, 28, 29, 33, 73, 74, 75, 77, 79, 82, 84, 88
Brush Lake, 26, 78, 86
Buccella
  depressa, 20, 21, 78, 98, 99
  parkerae, 78
Bulimina
  corrugata, 55
  gonzalezi, 37, 76, 96, 97
  reussi v. *navarroensis*, 76
  versa, 76

Cabo Negro, 19
Caldenius, 17
California, 84
Camacho, 64
Cameronian Stage, 3, 16, 29, 31, 55, 56, 79, 85, 86, 87, 90
Campanian-Cenomanian, 15
Camptonectes sp., 46
Candeina cecioni, 37, 83, 100, 101
Canelos River, 77
Canelos Sur River, 37, 39

Cape Horn, 9
Capitan Aracena Island, 9
Carbonaceous matter, 21, 61
Carmen area, 21, 27
Cassidulina
  cf. *brocha*, 21, 84, 100, 101
  rotulita, 84
Catalina, 23
Catalina no. 1 well, 86
Cenomanian, 3, 38, 82
Cerro Baquales, 17
Cerro Colo Colo, 15
Cerro Diadema conglomerate, 15
Cerro Donoso plutonic bodies, 17
Cerro Frailes, 17
Cerro Laurita, Brunswick Peninsula, 79
Cerro Paine, 17
Chañarcillo no. 4 well, 74
Chorrilla García, 71, 77
Ciaike no. 1 well, 89, 91
Cibicides cf. *djaffaensis*, 43, 84, 100, 101
Cibicidoides semiumbilicatulus 38, 88, 102, 103
Ciervo River Valley, 28, 31
Ciervos Formation, 64
Cisne no. 1 well, 26, 27, 28, 55
Cladophlebis, 10
Clarence Island, 9
Clarencia no. 1A well, 19, 22
Clarencian Stage, 3, 16, 29, 33, 70, 74, 91
Clavulina, 71
Clavulinoides
  chileana, 70
  rugulosa, 71
  szaboi, 70
Continental coal-bearing beds, 16
Cordillera de la Costa, 9
Cordillera Pinto, 17
Cordillera Vidal, 17, 23
Cormoran no. 1 well, 12, 52
Correlation with other areas, 53
Cretaceous, 38, 46, 48, 63, 78, 85
  Tertiary boundary, 15, 37, 64
  trough migration, 16
Cribroelphidium cf. *strattoni*, 79
Cristellaria
  asperuliformis, 74
  biexcuvata, 74
  stilla, 73
  tricarinella, 74
Cruceros
  area, 21, 28
  no. 1 well, 21, 27, 31
Cullen

  area, 55
  no. 6 well, 74
  no. 64 well, 48, 50
  -Tres Lagos area, 12
Cushman, 72, 74
Cyclammina cancellata, 21, 55, 75, 94, 95

Daly Creek, 27, 28
Danian, 3, 37
Darwin Cordillera, 9
  intrusives, 12, 16
Dawson Island, 15, 17
Dentalina soluta, 46
Denton Formation, Texas, 78
Diatoms, 61
Dioritic rocks, 9, 15
Discorbina semiumbilicata, 88
Discorbis minima, 43, 78, 96, 97
Discordia River, 43, 77, 85
Divisaderian Stage, 3, 17, 19, 23
Dorothia
  brevis, 72
  inflata, 72
  mordojovichi, 43, 71, 94, 95
  plummeri, 72
  principensis, 29, 72, 94, 95
Dungeness no. 1 well, 12, 52
Dungeness Point, 23

Eagles Ford shale, 80
East Province, 26
Echinoid spines, 20
El Ganzo River, 69
Elphidium
  aquafrescaensis, 37, 79, 98, 99
  crispum, 19
  patagonicum, 29, 79, 98, 99
  skyringensis, 79, 98, 99
El Salto no. 1 well, 24, 29, 32
ENAP P-7 well, 72, 79, 88
Eocene
  of Cuba, 72
  Krayenhagen shale, 72
  lower, 3, 29
  middle, 29
  in Tampico, 74
  in Trinidad, 82
  upper, 3, 29, 72
Epistomina porcellanea, 89
Epistominella texana 37, 78, 98, 99
Equisetites, 10
Esperanzian Stage, 3, 13, 46, 55, 68, 70, 72, 74, 75, 87, 91
Espora

no. 1 well, 83
Point, 23
Estancia
  El Salto, 21, 26
  Filaret, 25
  Gaviota, 21, 26
  Leonardo, 10
  Nueva no. 1 well, 80, 82
Estheria fauna, 10
Estratos con Favrella Formation, 64, 66
Estrecho de Magallanes, 22
Eugeosynclinal realm, 13
European stages, 6
Euxinic conditions, 13, 17
Evan no. 1 well, 13, 56
  to Vania no. 1 well section, 55
Extra-Andean Patagonia, 13, 17

Favrella
  americana, 46
  steinmanni, 46
Filaret no. 1 well, 21, 26
Fish remains, 61
Flora, 48
Florilus
  cf. *boueanus*, 20, 85, 100, 101
  scaphus, 20, 86, 100, 101
Fluvioglacial, 11
Fontaine River area, 46, 48, 49
Foothill belt, 9, 15
Foraminifera, 3, 12
  from Agua Fresca formation, 63
  from Brunswick Peninsula, 63
  charting, 65, 66
  chart, Sombrero no. 1 well, 61
  Upper Cretaceous-Tertiary sequence, 63
Foraminiferal species, 67
Foraminiferida, 67
Fuentes Bay, 38
Fundo San Jorge, 36, 37
Fusilinid-bearing limestone, 12

Gaudryinella, 71
  alexander, 71
Gault of Folkstone and Biggleswade, England, 70
Gaviota Lake, 20, 28
Gaviotian Stage, 3, 16, 20, 21, 26, 28, 55, 68, 76, 77, 78, 83, 84, 86, 87, 90
Gaviotian trough, 16
Germanian Stage, 3, 15, 26, 28, 37, 69, 71, 76, 77
Glauconite, 20, 21, 28, 61
Gleichenites cf. *san martini* Halle, 46

Globanomalina, 82
Globigerina
  (*G.*) *bulloides* Volger subsp. *naussi* Gandolfi, 81
  cretacea, 38, 82, 100, 101
  cf. *cretacea*, 80
  paradubia, 83
  triloculinoides, 29, 83, 100, 101
  wenzeli, 38, 83, 100, 101
Globigerinella escheri escheri, 81
Globoquadrina altispira globosa, 83
Globorotalia cf. *crassata* var. *aequa*, 29, 82
Globotruncana
  aspera, 81
  chileana, 38, 80, 81, 98, 99
  lapparenti, 82
  lapparenti var. subsp. *tricarinata*, 81, 98, 99
  marginata, 38, 81, 98, 99
Gneiss, 12, 48
Granites, 9
Granodiorite, 48
  gneiss, 48
Gravitites, 15, 38
Groove casts, 38
Gryphaea sp., 46
Gulf of Penas, 12
Gümbelina moremani, 80
Gyroidina
  infrafosa, 37, 87, 102, 103
  soldanii, 21, 28, 87, 102, 103

Haplophragmium inconstans var. *erectum*, 46
Hasterigerina, 82
  escheri escheri, 43, 81, 100, 101
  iota, 29, 82, 100, 101
Hauterivian, 3, 46
Hedbergella planispira, 43, 80, 98, 99
Herterohelix
  globulosa, 38, 80, 98, 99
  moremani, 43, 80, 98, 99
Hoeglundina
  elegans, 21, 28, 88, 102, 103
  eocenica, 89
  porcellanea, 46, 89, 104–105
Hollister, J. S., 63
Holocene, 17, 70
Holotypes, 67
Hoste Island, 9
Hoste Navarine Island, 13

Ignimbrites, 12
Inoceramus, 38, 61
Intra-Lazian-Peninsulian movements, 15

Itasca, Texas, 80

Japan, 86
Josefina no. 1 well, 68
Juan Mazio Peninsula, 19, 22
Jurassic,
  Middle, 12
  Upper, Lower Cretaceous, 46
  Upper, Oxfordian-Kimmeridgian, 46

Kaiatan, 82
Karreriella
  cushmani, 21, 28, 72, 94, 95
  cylindrica, 73
  marina, 72
Kerber no. 1 well, 37, 39, 55
Kniker, H. T., 63, 69, 73
Kreyenhagen shale, 72

Laboratory procedure, 65
La Golondrina, 10
Lagoonal conditions, 21
La Juanita, 10, 51
Lake
  Buenos Aires, 10
  Nahuelhuapi, 11
  Toro, 42, 43
La Leona, 10
La Matilda, 10
La Modesta, 10
La Peninsula, 42, 43
Laramian orogeny, 16
Lazian, lower, 71
Lazian-Peninsulian movements, 15
Lazian Stage, 3, 15, 38, 41, 72, 73, 74, 75, 76, 77, 78, 80, 81, 83, 84, 87, 88
Leña Dura no. 1 well, 68
Leña Dura River, 28, 29, 31, 32, 70
Lenticulina
  cf. *asperuliformis*, 29, 74, 96, 97
  besairiei, 46
  biexcavata, 46, 74, 96, 97
  reyesi, 43, 74, 96, 97
Liassic continental beds, 10
Lizard Springs formation, 85
Lower Barremian, 43
Lower Cretaceous, 13, 14, 38, 43, 46, 78
  basin, 15
  sedimentary rocks, 63
Lower Eocene-middle Miocene, 29
Lower Paleozoic series, 52
Lower Tertiary
  beds, 16
  deposits, 12, 21

Lucina cf. *neugoensis*, 46

Maach, R., 14
MacPhearson hill, 20
MacPhearsonian Stage, 3, 16, 17, 19, 20, 24, 78, 86
Madagascar, 13
Madre de Dios Basin, 12, 51
Maestricktian, 3, 38, 64
Magallanes Basin, 3, 5, 10, 11, 12, 13, 16, 53, 55, 56, 66, 67, 68, 69, 72, 73, 83, 86
  Foraminifera, 59
  geosyncline, 13
  microfauna, 59, 65
  province, 9, 11, 15, 16
  trough, 13, 56
Magallanian Steppe, 9
Main Cordillera, 9, 11, 12, 13, 15, 16, 17, 56
Mal Paso formation, 69
Manatiales no. 1 well, 84
Manzanian Stage, 3, 16, 29, 35, 36, 37, 56, 69, 72, 73, 79, 83, 87, 88
Manzano
  area, 29
  no. 1 well, 69, 72
  no. 5 well, 28, 30
  no. 7 well, 16, 21, 26, 33, 34, 36, 37, 82, 83, 85, 90, 91
Margas Formation, 90
Marginulina Knikerae, 28, 75, 96, 97
Marginulinopsis lituoides, 46
Maria Emilia
  no. 2 well, 12, 48, 51, 52, 55
  no. 3 well, 48, 50
Marine
  deposits, 16
  flysch, 15
Marks, Vienna Basin, 85
Martinez, R., 63
Mazian Stage, 3, 17, 19, 22
Megafauna, 46
Mellizos no. 1 well, 78
Metamorphic
  basement, 48
  rocks, 12
Microfaunal examination, 65
Microflora, 10
Minas River Valley, 21, 27, 28, 30
Miocene, 3
Miocene-Pliocene tuff, 11
Miogeanticlinal ridge, 13
Miogeosyncline realm, 13

Mirador Hills, 21, 28
Miradorian Stage, 3, 16, 21, 27, 28, 55, 68, 70, 73, 76, 77, 78, 83, 84, 86, 87, 90
Miraflores wells, 56
Mirante Martinez, 52
Mobile belt, 11
Molasse type, 15
Mollusks, 20, 61
Monte Aymond no. 2 well, 84
Monte Buckland-Fiordo, 48
Monte Tres Picos, 12
Monton Point, 48, 49
Monzonite, 15
Mordojovich C., 63
Moritzian Stage, 3, 29, 32, 56, 72, 75, 79, 85, 86, 87, 88, 90
Mount Luis de Soboya, 9
Museo Nacional de Historia, Santiago, Chile, 67, 77

Natland-Gonzalez system, 64
Nautilus
 crepidula, 73
 scapha, 86
Navarino Island, 17
Neocomian, 13, 14
 beds, 73
New Zealand, 82
N. Lubecka, 51
Nonion, 82
 boueanum, 85
 iota, 82
 scaphum, 86
Nonionella
 auris, 86, 100, 101
 miocenica v. *stella*, 86
 modesta, 86
 pauciloba, 86
 pseudo-auris, 86
 pulchella, 20, 86, 100, 101
Nonionina
 boueana, 85
 bulloides, 86
 escheri, 81
Nonmarine, 16
Notobatrachus, 10

Oazian Stage, 3, 15, 16, 36, 37, 56, 69, 76, 78, 87
Oazian to Rosarian Stages, 16
Oil reservoirs, 56
Oligocene
 lower, 3, 28
 upper, 3

Ophiolites, 9
Osangularia
 brunswickensis, 88, 102, 103
 lens, 88
 plummerae, 88
 tenuis carinata, 88
Ostracods, 20, 61
Otozamites, 10
 sanchtaecrucis, 46
Oxfordian, 3, 12, 13, 46
Oxfordian-Kimmeridgian, 3, 13, 46

Pacific drainage system, 17
Paleocene, 3, 16, 37
 fauna, 64
Paleocurrents, 15
Paleoecologic summary, 55
Paleoecology, 46, 56
Paleoslope, 15
Paleozoic
 basement, 10
 granodiorite, 12
 Jurassic basin, 10
 lower, 12
 rock outcrops, 51
 series, 48
 time, 10
Pampa Larga
 area, 68
 no. 1A well, 24, 29, 32, 38, 40, 41, 42, 43, 44, 45, 46, 47, 48, 49, 68, 70, 71, 72, 90
Paraguaya no. 1 well, 78
Paratypes, 67
Parker and Jones, 70
Patagonia, 63
Patagonian Andes, 9, 11
Patagonian Archipelago, 15
Patagonian Cordillera, 16, 17
Peninsula Brunswick, 32
Peninsula Buckland, 52
Peninsula Espora, 12
Peninsulian-Pratian Stages, 14, 43
Peninsulian Stage, 38, 42, 43, 55, 68, 70, 71, 72, 73, 74, 80, 81, 82, 83, 85, 91
Penitente River, 23
Pennsylvanian
 lower, 12
Pericratonic Basin, 13
Permian
 granite, 10
 lower, 12
Petroleum, 48
Piedra sholte, 10

Planktonic Foraminifera, 66
Planulina popenoei, 38, 84, 100, 101
Plectina
   elongata, 29, 72, 94, 95
   eocenica, 72
   garzaensis, 72
Pleistocene, 3, 19
   glaciation, 17
   –Holocene time, 17
Pliocene
   epoch, 17
   lower, 3
Plutonic bodies, 17
Polymorphina martinezi, 46, 75, 96, 97
Polystomella strattoni, 79
Portlandian (Tithonian) time, 3, 13, 46
Posesión no. 1 well, 12, 52, 87
Postorogenic period, 15, 16
Praeglobobulimina
   kickapooensis, 37, 77, 96, 97
   pupoides, 20, 77, 96, 97
   subcalva, 55
Prat Point, 69, 78
Pratian Stage, 3, 13, 43, 73, 74, 91
Precambrian, 12
Previous work, 63
Psamminopelta, 68
   minima, 38, 67, 94, 95
   venezuelana, 21, 28, 67, 94, 95
Puerto Nuevo, 76
Pullenia
   bulloides, 21, 28, 55, 102, 103
   cretacea, 87
   erecta, 87
   natlandi, 43, 87, 102, 103
Pulvinulinella texana, 78
Punta Arenas, 21, 22, 30, 51, 63, 64
Punta Baja, 52
Punta del Cerro no. 1 well, 23, 27, 28, 29, 30, 31, 32, 33, 91
Punta Delgada area, 12

Quaternary, 19
   glaciation, 11

Radiolaria, 55, 61, 63
   Nassellaria sp., 3, 46, 91, 104, 105
   Spumellaria? sp. 1, 29, 43, 90, 91, 104, 105
   sp. 2, 43, 90, 91
   sp. 2, 43, 90, 91
   sp. 4, 43, 90, 104, 105
   sp. 5, 29, 91, 104, 105
   sp. 6, 46, 91, 104, 105

Rectuvigerina ongleyi, 28, 77, 96, 97
Reinholdella
   fuenzalidai, 46, 89, 104, 105
   cf. *quadrilocula*, 13, 46, 89, 104, 105
Rhyolitic
   flows, 12
   rocks, 10
Riescoian
   Stage, 3, 15, 16, 17, 38, 40, 55, 67, 69, 73, 76, 77, 78, 80, 83, 85, 87
   time, 15
Riesco Island, 16, 17, 38, 40, 69, 76, 87
Rinconian
   sequence, 13
   Stage, 3, 13, 15, 16, 46, 48, 55, 73, 74, 75, 84, 89, 91
Rinconian-Divisidarian, 17
Ronconian-Riescoian, 16
Rincon River, 46, 47, 48, 49, 50
Rio Chico
   area, 21, 27, 28
   no. 1 well, 56
Rio Chico Arch, 10, 11, 12
Rio del Oro no. 1 well, 17, 55, 90
Rio Minas, 30
Robertina arctica, 20, 90, 104, 105
Rocallosa Formation, 15, 63, 64
Rocallosa Point, 38, 40, 69, 76, 87
Rosalina marginata, 81
Rosarian Stage, 3, 16, 28, 30, 55, 77, 79
Rosario Creek, 28, 30
Rzehkina, 68

Sample
   collecting, 61
   examination, 65, 66
   plotting, 65, 66
   plotting header, 65, 66
   washing, 65
San Antonio no. 1 well, 21, 26, 29, 32, 70, 79
San Diego, 56
San Joaquin Valley, 84
San Sebastián no. 1 well, 80, 81, 83, 84, 88
Santa Cruz Basin, 9, 51
Santa Cruz Province, 10, 11, 12, 15, 16, 17, 56
Santa Cruz River, 9
Santa Inez Island, 9
Santa Maria River, 34, 75
   valley, 29, 54
Santonian, 3, 38
Schist, 48
Scripps Institution of Oceanography, 56

Sebastinian Stage, 3, 20, 25
Seewer kalk, 82
Seismic data, 11
Sena Poco Esperanza, 13
Seno Alimirantazgo, 48, 49
Seno Contral, 52
Seno Otway, 69, 73, 77, 78, 84
Sialo-simatic suite, 15
Siegfus, Stanley S., 72, 74
Siphogenerina ongleyi, 77
Skyring Sound, 15, 17
Sofia conglomerate, 15
Sombrero
  area, 55
  no. 1 well, 26, 61, 66, 73
  no. 3 well, 74
South America-Africa, 13
Southern Chile, 63
Southern Patagonia (Chile), 3
Sphaeroidina bulloides, 21, 28, 75, 96, 97
Sphenopteris patagonica, 46
Spiroplectammina
  adamsi, 29, 68, 94, 95
  brunswickensis, 37, 69, 94, 95
  grzybowskii, 37, 69, 94, 95
  gutierrezi, 38, 69, 94, 95
Spiroplectinata annectens, 43, 70, 71
Spirosigmoilinella, 67
Spongastericus, 91
Springhill
  formation, 13, 64, 66
  platform, 11, 13, 15
  sands, 48
Stipanicic, 10
Stratigraphic and paleontologic conditions, 63
Suero, 10

Tampico Embayment, 88
Taylor marl, 78
Temblor Formation, 78
Tenerife Hill, 43, 44, 45, 46, 47
Tenerifian Stage, 3, 14, 43, 46, 55, 69, 78, 84, 87, 90
Tepuel, 51
Tertiary, 19
  Basin, 56
  sedimentary and microfaunal sequence, 64
Textularia
  annectans, 70
  globulosa, 80
Textulariina, 67
Tierra del Fuego, 9, 11, 12, 15, 16, 17, 19, 28, 29, 33, 48, 49, 63, 64, 65, 73, 76, 77, 78, 85, 86, 87

Tithonian-Neocomian, 13
Tobifera
  relief, 12
  series, 11, 12, 13, 48, 50
Todd, R., 69
Tonalite, 15
Toro Hill, 38, 41
Toro Lake, 42
Toro no. 2 well, 43
Trachydolerite, 17
Traction transport, 56
Tranquilo no. 2 well, 37, 39
Tres Brazos formation, 16
Tres Brazos no. 1 well, 35, 36, 37
Tres Brazos River, 29, 33, 34, 74
Triassic
  continental beds, 10, 12
  in southern Patagonia, 12
Trifarina angulosa, 20, 77, 96, 97
Trigonia, sp., 46
Trinidad, 82, 85
Tritaxia, 70
  chileana, 29, 70, 94, 95
  porteri, 43, 71, 94, 95
  rugulosa, 37, 71, 94, 95
Trochammina cf. *inflata*, 21, 28, 70, 94, 95
Tuffs, 12
Turbidite
  sand, 56
  sequence, 38
Turbidity currents, 38, 56
Turonian-Senonian, 82

Ultima Esperanza, 12, 13, 15, 16, 17, 24, 33, 38, 40, 41, 42, 43, 44, 45, 46, 47, 48, 49
Upper Cretaceous, 38, 69, 84
  basin, 15
  and lower Tertiary deposits, 12
  sandstone, 56
  time, 11, 12, 13
Upper Eocene, 29
Upper Jurassic basin, 13
Upper Jurassic-Cretaceous time, 11, 12, 13
Upper Miocene, 19
Upper Paleozoic, 12
Upper Permian, 12
U.S. National Museum, Washington, D.C., 67-71
Uvigerina angulosa, 77

Vaginulinopsis ectypa, 46
Valdes conglomerate, 15
Valvulina auris, 86

Vania no. 1 well, 38, 40, 46, 48, 49, 55, 61, 66
Velasco shale, 88
Victoria Norte no. 2 well, 75
Victoria Sur no. 1 well, 68
Virgulinella severini, 28, 85, 100, 101
Volcanic
  activity, 17
  dust, 6
  ejecta, 61
  glass, 19

West province, 26, 28
White chalk, England, 82

Yahgan sediments, 14
Yazoo clay, Mississippi, 72

WITHDRAWN